T0309309

THE INTIMATE LIFE OF COMPUTERS

THE INTIMATE LIFE
OF COMPUTERS

Digitizing Domesticity in the 1980s

REEM HILU

University of Minnesota Press

Minneapolis

London

A different version of chapter 2 was published as "Calculating Couples: Computing Intimacy and 1980s Romance Software," *Camera Obscura* 38, no. 2 (September 2023): 145–71, https://doi.org/10.1215/02705346-10654941; copyright Duke University Press. An earlier version of chapter 3 was published as "Girl Talk and Girl Tech: Computer Talking Dolls and the Sounds of Girls' Play," *The Velvet Light Trap* 78 (Fall 2016): 4–21; copyright 2016 by the University of Texas Press; all rights reserved.

Published by the University of Minnesota Press
111 Third Avenue South, Suite 290
Minneapolis, MN 55401-2520
http://www.upress.umn.edu

ISBN 978-1-5179-1664-0 (hc)
ISBN 978-1-5179-1665-7 (pb)

Library of Congress record available at https://lccn.loc.gov/2024024842

CONTENTS

Introduction: Defining Companionate Computing   1

1 A Ménage à Trois with Your Computer: Romance Software as Mediator for Couples   29

2 "Not an Appliance, but a Friend": Personal Robots and Participant Fatherhood   63

3 "A Doll That Understands You": Computer Talking Dolls as Parenting Proxies   99

4 Sex and the Singles Game: Adult Games, Cringe, and Critiques of Masculine Seduction   131

Coda: Companionate Computing and Its Echoes   171

Acknowledgments   181

Notes   185

Index   213

INTRODUCTION

Defining Companionate Computing

FOR THE 2022 SUPER BOWL, Amazon produced a commercial showing celebrity couple Scarlett Johansson and Colin Jost interacting with the Alexa-enabled Amazon Echo.[1] After marveling at the Echo's ability to prepare their home to their liking for game day, the pair comment that it is like the virtual voice assistant can read minds. The rest of the commercial offers speculative examples of what would happen if Alexa did have that ability. The couple wakes up in bed and Alexa, intuiting Jost's displeasure with his wife's morning breath, announces an order of fresh mint mouthwash. Next, they are in the kitchen. Johansson listens to her husband's endless chatter and Alexa, reading her irritation, turns on the blender to drown him out. The commercial proceeds to show a few more examples in which Alexa exposes harmless lies that the couple may use to keep their relationship running more smoothly or to spare each other's feelings. In all these instances, Alexa's actions or announcements reveal minor grievances that are best left hidden, at least from the other member of the couple in this shared household. At the conclusion of the commercial, Johansson and Jost agree it is best Alexa cannot read minds. Typical of many Amazon commercials that portray instances in which the virtual assistant acts out, the thought experiment has shown that Alexa is perfect just the way it is—a helpful companion, but one that enables rather than impedes domestic relationships.

According to Amazon, Alexa is the perfect companion because it is calibrated to be unobtrusive and to ease relationships between family members rather than disrupt them. In an interview describing Amazon's investment in technologies like the Echo, head of devices and services David Limp explains this vision of the role that digital technologies should take up in domestic spaces. Citing his own family as an example, Limp claims that the motivation behind this approach to computing is to combat the atomization that can sometimes result, even within a family, from the spread of digital devices in the home. According to Limp, using Amazon's technology "allows you to pick up your head and enjoy moments as a family. . . . So, when you want to play music, instead of putting those earphones in, you're actually saying 'play music' and enjoying it together. When you're watching TV, you can say 'tune to ESPN,' and all watch the sporting event together and be one as a family. And then when it disappears, it goes away, it fades into the background and the technology is not overwhelming."[2] In Limp's formulation, technologies like Alexa are desirable not necessarily due to any specific content they provide, but because of how they channel and improve relationships between family members.

Amazon sells a fantasy in which digital media serve as companions but also as tools that foster companionship with others in the home. The company implies that its smart technologies are uniquely capable of facilitating family relations. In fact, this approach to computers and family life has a much longer and contested history. Long before the Echo, efforts were made by different producers of hardware and software in the 1980s to integrate then novel computers into the domestic cultures of the home. These parties were influenced by the context of contemporaneous companionate relationships as they developed applications for computing that could potentially augment and remediate familial and intimate bonds. To unpack the foundations of this longer history, *The Intimate Life of Computers* examines the spread of computing into domestic spaces in the United States in the 1980s, focusing on those applications that encouraged users to conduct companionate relationships with and through computers.

I call this a history of "companionate computing" rather than "home computing," "personal computing," or "domestic computing" to highlight not just how computers came to be adopted into the home in the 1980s, but to speak more explicitly to how they were envisioned as media to help negotiate and bolster companionate relationships. In doing so, this book also emphasizes the impact of cultural shifts in family relationships on the development of computing technologies.

Granted, when computers became available for the home, they were adopted first by hobbyists, who were mostly men; when the technology spread to more general consumers, the market for computing was still conceived of as largely masculine. Fears circulated that computers would function to isolate men, even at home. This was the case in the 1980s when commentators and academics like Sherry Turkle spoke of hackers and hobbyists choosing computers over personal relationships. But this impression of the development of computing continues in more contemporary studies that trace the early appeal of computers to their ability to cater to fantasies of atomized and powerful individual experiences.[3]

Although women were not often the creators nor even the primary intended audience for many of the computer technologies and programs created to be used in homes in this period, some in the industry nonetheless considered how computers at home would affect both women and the whole family. Some producers were contending with middle-class domestic culture, feminist critique, and their perceptions of how women would respond to computing in order to inform the development of software and hardware that aimed not to isolate men, but to enhance companionate family relations.

This consideration of women and family produced a way of thinking of computers as technologies of domestic relationality, and the resulting applications reveal the influence of domestic histories and cultures in shaping computer history. In the interest of uncovering how women's and feminist culture shapes the development of computing, even if indirectly, this book explores these companionate applications and the contexts in which they were produced. In its attention to these influences, the approach in this book is inspired by previous feminist media histories that have emphasized the significance of domesticity and the companionate family in shaping media technologies like radio, television, and phonography.[4]

Reaching a point when computing technology could be understood as unobtrusive or even beneficial to the home and family was not uncontested or inevitable. Although the late 1970s and 1980s would have been the first time that computers spread in significant numbers into U.S. homes, newspaper and magazine commentary produced at the time expressed some uncertainty about how to make sense of what was often referred to as "home computing."

In some cases, popular and industry press used the term "home computer" simply as a commercial category to distinguish lower-end "home computers" like the Commodore VIC-20 or TI-99/4A from high-cost and more technologically powerful machines like the IBM PC or Apple II that were more associated with business or specialist uses.[5] Aside from its use in classifying hardware, there were also contestations in the press about whether "home computing" described something distinct or was just a term for computers operated at home. As late as 1985, John Sculley, then president of Apple Computer, even said of the home computer market, "It doesn't exist. People use computers in the home, of course, but for education and running a small business. There are not uses *in* the home itself." Voicing agreement, Trip Hawkins, founder of the major software company Electronic Arts, claimed, "The primary use for home computers is still entertainment. No one has yet figured how to make the computer part of the social fabric of the home."[6] Statements like these cultivated a distinction that was circulating more generally in commentary about computing: on one hand were existing popular uses of computers for games or professional applications that just happened to be done in the home, and on the other hand was a yet unrealized and potentially misguided version of home computing more integrated with feminized domestic space.

This uncertainty about home computing was not limited to hardware and software executives, but also repeated by journalists. One article suggested, "Few households amass the volume of data that justifies a computer." Another argued, "It may take more discipline or imagination, or more sophisticated needs than most of us have to weave the computer into the fabric of our daily lives."[7]

One of the focal points for this contestation over computers' assimilation into the "fabric of the home" involved questions about how they might affect domestic and familial relationships. Immediately upon their arrival, some feared that home computers would disrupt existing relationships. Even as early as 1975, when the only devices available for the home were kit computers that required specialist knowledge to assemble and operate, the hobbyist magazine *Byte* devoted its December issue cover to a tongue-in-cheek speculation about the kind of family discord that computers in the home might cause (Figure 1). In this cartoon image, a man holds his children back and looks wildly at a new computer under the Christmas tree as his angry wife stares daggers at Santa Claus, who has delivered this "gift" into their home. As the editors explain, the cartoon image on their cover "illustrates the impact of these new toys upon traditional relationships."[8]

Concerns about computers and domestic affairs continued as computers spread beyond hobbyist users in the 1980s. One of the most common anxieties came in the form of discourses about computer widows who had lost their husbands to the home computer.[9] Apprehensions about computers as part of family life came in other forms as well: psychologists spoke of patients who could no longer communicate effectively with family after adapting to the rational style of interaction they used with computers, studies suggested that parents were favoring their sons when spending on computers and putting their daughters at a disadvantage, and critics feared that computers might seduce men away from the challenges of more complex interpersonal encounters.[10]

Figure 1. *Byte* magazine cover from December 1975 speculating on the computer's effect on family relationships.

Amid such misgivings about the effect of computing on family relations, some 1980s software and hardware producers were using their understanding of domestic life to craft uses for computing that might enhance rather than undermine companionate family relations. Lori Reed has argued that the management of anxieties about computers and family were part of the rearticulation of the computer from a "cold, distant, and feared military machine, or a tool used in isolation by a socially dysfunctional male" to "socially-friendly and family-friendly machines."[11] She discusses how this was achieved by "configuring users," managing their computer phobia and philia, and teaching them the proper and improper ways to use the devices as part of family life.

In my own approach, I am concerned instead with how these anxieties about relationships shaped the design of computing hardware and software products that attempted to integrate computers into family life and imagined computers as a relational, social, or interpersonal medium— that is, a medium for domestic relationality. These computing applications were intended to bolster the heteronormative family in the face of the unique challenges it was facing in the 1980s. But they also had the potential to generate unexpected attachments or highlight contradictions in ideals of family connection and intimacy. To get at the ambivalence that resulted when hardware and software makers envisioned computers as technologies for domestic relationality, this book engages in close readings of specific computer programs and games as well as hardware like robots and microprocessor-enabled dolls and analyzes how they were designed to shape and channel relationships.

Computers were mobilized as a medium that encouraged and conducted specific kinds of companionate sociality within the home, but what it looked like for computers to be used as media for domestic relations in the 1980s varied. As the chapters in this book will discuss, software and hardware makers produced programs that mediated between coupled users and guided them in modes of more effective romantic and sexual interaction while they used the computer together; developed cute, pet-like robots that fathers could program with their families to extend and support their performances of participant fatherhood; manufactured microprocessor-embedded dolls that imitated conversations and served as proxies for parental attention for the girls who played with them; and even designed games that playfully simulated heterosexual seduction and encouraged users to critique and reflect on practices of heteromasculinity.

These computing products were informed by longer histories of domestic practices and cultures as well as changing popular expectations placed on middle-class families and gender relations in the 1980s. They engaged with shifting ideals for gender relations, coupled intimacy, and childcare and also relied on long-standing practices such as doll play, domestic appliance design, and therapeutic and self-help culture. That influence is apparent both in the products created and the promotions used to sell and make sense of them. The result of hardware and software makers contending with domesticity was not just new applications for computers, but a way of understanding the technology as a kind of social medium—one that mediated relationships among family members and drew on existing practices and cultures associated with home and family life.

Accounts of the influence of domesticity and companionate relationships in shaping histories of computing are still largely absent, both in popular and academic accounts of computing in the 1980s. Popular treatments do not often characterize computing in this period as particularly domestic. Narratives like Michael Swaine and Paul Freiberger's *Fire in the Valley* and Steven Levy's *Hackers* recount the stories of hobbyists and entrepreneurs who delivered computing to the masses by developing software products like spreadsheets, word processors, database programs, and especially games—applications they source to the traditions of hackers, hobbyists, and resourceful entrepreneurs rather than to any domestic precursors.[12]

Despite being the decade in which computing was first adopted into millions of homes, the 1980s has not received the same sustained academic focus as other periods in computer history. Of course, some scholars have produced insightful accounts of how during this period computers were refigured from hobbyist machines to everyday household consumer products. Lori Emerson and Michael L. Black, for example, have both discussed the rise of user-friendly "appliance computing" in the late 1970s and 1980s. Black argues that computers like the Commodore PET and TRS-80 leveraged the rhetorics of user-friendliness and universal usability to adapt computers to a nonspecialist home market. For Black, this focus on usability was executed not only in promotional material but most significantly in the design of these computers as black boxes or "appliances" that neither encouraged expansion nor required tinkering to use.[13] Emerson and Black provide useful insight into the politics of user-friendly design and help demonstrate how it contributed to the spread of computing. As Emerson argues, in the guise of user-friendliness and transparent interfaces, this model of computing "further alienates the user from having access to the underlying workings of the device" and "turns all computing devices into appliances for the consumption of content instead of multifunctional, generative devices for reading as well as writing or producing content."[14] Still, these accounts are most concerned with the way home computers were modeled as easy to use like domestic appliances, rather than with the particular domestic uses that were attributed to them or the ways they could be integrated into family life.

Most recently, Laine Nooney's work on the history of the Apple II has addressed the development of a uniquely home-based mode of computing. Like Black, Nooney argues that personal computing had to be invented rather than just gifted to ordinary users by the genius of hobbyists. Her chapter on the popular desktop publishing program The Print Shop shows one way this was achieved specifically for the home. Nooney argues that The Print Shop innovated by demonstrating how computers could find a place in people's everyday lives. As she argues, "home computing was more than simply a software category; it embodied an ethos of living with computers, of imagining one's *life* (and not just one's home) as available to be computerized."[15] Although this account addresses computing as part of the everyday life of the home, it does not address how specific aspects of domestic practice influenced the form that home computing took or how computers were mobilized as tools to mediate the pressures on companionate relationships.

I use the term companionate computing to foreground this book's distinct interest in the influence of domestic culture and companionate relationships on the development of computing in the home and to indicate a focus on applications of computing mobilized to support and mediate domestic relationships. I use this term to emphasize a focus of inquiry, but also to distinguish the approach in this book from other models of delimiting this history. Companionate computing is useful strategically even though it was not used to describe computing in the 1980s the same way that home computing or personal computing were.

When commentators and scholars discuss home or personal computing as it existed in the 1980s, they are often referring to desktop or microcomputers and discussing these general-purpose machines. As described in *Computer: A History of the Information Machine*, it is the personal computer as "the configuration of a self-contained machine, somewhat like a typewriter, with a keyboard and screen, an internal microprocessor-based computing engine, and a floppy disk for long-term data storage."[16] Developments in microprocessor technology in the 1970s enabled the creation of desktop microcomputers, making computing power more accessible. At the same time, microprocessors spread into the home in other forms. Video game consoles are likely the other most prominent example of this spread; but as James Hay has documented, microprocessors also proliferated in the form of what he calls programmable technologies. Microprocessors could be found embedded in appliances like microwaves, coffee makers, and exercise machines.[17] Because the term companionate computing is not already associated with a particular technological configuration, it allows the project to move across these technological differences to find a common mode of computing practice.

Conflating these different embodiments of computing, including microcomputers, software, games, robots, and talking dolls, might seem strange to some historians of technology. These products did not all arise from the same industry, at least narrowly defined, and they represent quite different technological configurations and interfaces. For example, it may be unexpected to classify appliances or toys fitted with microprocessors as computers. Robots and microprocessor-embedded talking dolls often lacked keyboards or screens, which were expected with desktop computers in the 1980s (even if monitors were sometimes sold separately). In the case of talking dolls, they had limited and dedicated uses, meaning that users could not easily program them. As a result, they departed from some expectations and fantasies commonly associated with microcomputers. But computer historians have demonstrated that the vision of personal computing as stand-alone commodity devices competed with many others prior to the spread of microcomputing; many envisioned computing more as a public utility accessed through remote personal terminals.[18] Similarly, programmable desktop computers represent only part of the story of how computing entered everyday life in the 1980s.

To better account for the variety of ways that computer technology was mobilized as part of companionate relationality, I have had to expand my perspective. This broadened view includes examples beyond desktop microcomputers. It is key to acknowledge other modes of computer interaction, especially when they present alternative forms of relating to the digital. The examples discussed in *The Intimate Life of Computers* are not all representative of programming or computing in the way hobbyists may have imagined it. Learning to program, entering numbers into spreadsheet software like VisiCalc, playing a flight simulator, talking to a doll that recognizes your voice using digital signal processing, setting an alarm on a robot using menu-based screens—these are all acts of computing.

The subject of this book is also distinct from "personal computing." That expression has developed a more abstract, expansive meaning in computer scholarship, to connote a mode of individualized interaction. Scholars like Thomas Streeter, Fred Turner, and Thierry Bardini have traced the concept of personal computing to a period long before the existence of commodity devices we often call personal computers.[19] Figures like J. C. R. Licklider and Douglas Engelbart discovered the "holding power" or "self-motivating exhilaration" of interacting with computers in the 1950s and 1960s. As one example, according to Streeter, while leading the computing projects of the U.S. Defense Advanced Research Projects Agency, Licklider imagined informal conversational exchanges with the computer, envisioning it more as a communication device than a calculating machine. Figures like Licklider helped link romantic tropes to the experience of computing.[20] Accounts like these are useful because they do not collapse personal computing onto a particular hardware configuration, application, or interface, but speak to a way of envisioning or interacting with computers as affectively and personally charged. They also suggest that the dream of personal computing did not originate only from hobbyists; rather, it developed earlier in military-funded projects and corporate initiatives.

Although this history is crucial to the developments that resulted in computers becoming available for adoption into the home, they describe a different mode of interacting with computers than those that fall under companionate computing. Accounts of personal computing tend to emphasize the power and appeal of experiences of individual mastery and control enabled by individual interactions with computers.

For example, when Streeter discusses personal computing in the 1980s, he focuses on the way personal computers functioned as socially evocative objects that made neoliberal and free market ideologies seem vivid and plausible. As he argues, "the experience of reading about, buying, and using microcomputers created a kind of congruence between an everyday life experience and the neoclassical economic vision—the vision of a world of isolated individuals operating apart, without dependence on others, individuals in a condition of self-mastery, rationally calculating prices and technology."[21] Streeter's concern in *The Net Effect* is how computers came to be related to the romantic self. Although Streeter acknowledges that this form of selfhood as abstract masculine individualism is not exhaustive, he sees it as the construction of selfhood that most influenced the trajectory of the reception of the internet in later decades. Streeter's work in connecting the history of computing to wider social contexts is extremely generative, but he is interested in microcomputers as "an ideological condensation symbol, for thinking about big social relationships."[22]

The Net Effect is focused on how computers helped us think about social relations, rather than how they mediated relationships in the family and home. The discussion of companionate computing in this book foregrounds a different facet of the cultural and social context of the 1980s: that of the home and the relationships occurring in middle-class family life. While Streeter helps us see how personal computing as it was constructed in the 1980s aligned with ideologies of neoliberal individualism, by shifting focus to companionate relationships I offer a different perspective on this neoliberal moment. Even when computers were used by multiple people together and not just individual users, the goal was to buttress the family, making it a better unit for private, neoliberal accumulation. My focus on family relationships emphasizes how computers were not just objects for thinking about relationships, but media through which users related to each other. Although the examples discussed in this book were not immediately influential in the way Streeter's were, they offer another thread of the history through which computers became intertwined with social relationships and were seen as communication technologies.

Furthermore, attending to the cultural context of family life and heterosexual relationships in defining computers helps uncover models for computing experience that did more than just support abstract masculine identity and mastery, which are often emphasized in histories of personal computing. As software and hardware makers developed applications of the computer for domestic use, they addressed users not as isolated, masculine subjects, but rather within their roles in the companionate family. Computer users were figured as fathers, daughters, and couples. As close readings will help demonstrate, the examples discussed in this book offered more various types of pleasures than that of mastery and control over a yielding computer—sometimes encouraging users to submit to the computer's commands or to treat it with care, and sometimes fostering feelings of anxiety and insecurity rather than dominance.

This book claims that some examples of 1980s computing—their development influenced by feminist contestations over companionate relations—served as technologies for domestic relationality, addressed a greater variety of users in their relational and domestic contexts, and offered forms of interaction that went beyond mastery and control. But I do not intend to reclaim this period or these products as socially progressive, feminist, or inclusive. Undoubtedly, scholars like Joy Lisi Rankin are right to critique the privatization of computing and the manner through which personal computers became exclusive commodities.[23] Although computers were newly accessible to private households beginning in the late 1970s, they were only available to the most privileged families. The applications of computing discussed in this book were prohibitively expensive for most Americans. By the end of the 1980s, when only 15 percent of all U.S. households owned computers, those in the top 20 percent of the population in terms of income were twice as likely to own computers than the average family and five times as likely as low-income families. In 1984, when the median U.S. income was $26,430, computers were most likely to be found in households with incomes of $50,000 or more.

Income and class were not the only determinants of computer uptake. Statistics show that in the 1980s, even in households that had computers, more men and boys used them than did women and girls.[24] As many have noted, women's representation in computer science professions began to decline in the mid-1980s even as women's participation in other science and technology fields increased. The unequal access of boys to computers in the home is often discussed as a contributing factor to this trend.[25]

Home computing was also unequally divided by race. By the end of the 1980s, less than 7 percent of Black households had computers, compared to the national average of 15 percent. Unequal access to computing in schools only added to gender, race, and class divides. Overall, data show that in the 1980s, home computers were most associated with white, middle- and upper middle-class homes, most often those headed by adults in professional or managerial occupations, and most commonly in the homes of families with children.[26] As computers were adopted into these professional homes, they were specifically adapted to accommodate expectations for these types of family arrangements.

Computers spread in an unequal manner in the 1980s and remained an exclusive technology throughout the decade. But this was still a fundamental moment in which computing contended with a new spatial and relational context as it started to address middle-class families in domestic spaces. The contestations over the technology's place in the home and how it was figured by some software and hardware makers as a technology to aid domestic relationships is crucial to understanding how computing came to be so deeply entwined in everyday life. This continued even as digital media came to be distributed more widely in later decades.

Bringing Computers Home

Currently, it is difficult to imagine middle-class American homes devoid of digital technologies. Especially in the past decade, domestic applications of computing and digital media have proliferated as a common part of family life. The Alexa digital voice assistant installed in the Echo smart speaker and other devices is a prominent example of this phenomenon. Amazon reported that as of 2019, more than a hundred million Alexa-equipped devices had been sold.[27] Amazon's other domestic products include the Echo Show, the Glow interactive video-calling and projector combination, and even a pet-like robot called Astro. But Amazon is neither the first, nor the only company successfully marketing digital devices for the home and family. Long before Alexa, iRobot—a company founded by MIT roboticists that spent its early years developing robotics for U.S. Defense and space exploration projects—launched the Roomba home vacuuming robot in 2002; as of 2020, iRobot has sold more than thirty million home robots.[28] iRobot and Amazon are joined by other companies like Google, Microsoft, and Sonos that offer their own brand of digital domestic assistance. Google's Nest technologies, for example, include smart thermostats, doorbells, and locks. Digital technologies are included in sundry other home applications including smart TVs, sound systems, and security devices. Although concerns remain about this spread of technology into intimate parts of everyday life, especially with regard to privacy, the commercial proliferation of these technologies indicates a widespread acceptance of computing's role in the domestic family space.

Yet, before the 1970s, it was nearly impossible to access computing technology at home.[29] Over the course of the 1950s and 1960s, the computer industry primarily served government offices, universities, and large corporations. UNIVAC, one of the first commercial digital computers in the United States, was installed at places like the U.S. Census Bureau, the Pentagon, and the offices of U.S. Steel, Dupont, General Electric, and Metropolitan Life, where it handled payroll, inventory control, and accounting. Other companies entered the mainframe computing market along with Remington Rand's UNIVAC, including General Electric, Honeywell, and IBM.[30] With their strong associations to the corporate world, in the public imagination mainframes most often represented anxieties about the automation of the workplace and society at large. But their domain did not extend into the home and appeared far from concerning companionate family relations.

In the 1960s, those mainframes were joined by smaller and more interactive minicomputers, like the PDP line of computers from Digital Equipment Corporation (DEC). Although called a minicomputer, the PDP-8, for example, was priced initially at $18,000 and weighed 250 pounds; clearly they were still institutional technologies.[31] Still, minicomputers allowed a wider spread of computing technology for government, academic, and corporate customers. Within these spaces, distanced from home and family, relationships with computers and conceptions of computer interactions were shifting for some users. Paul Ceruzzi argues that minicomputers were significant because they "succeeded in creating the *illusion* that each user was in fact being given the full attention and resources of the computer. That illusion created a mental model of what computing could be—to those fortunate enough to have access to one."[32]

As Fred Turner has documented, this period, too, saw computers become entangled with "New Communalist visions of consciousness and community" through figures like Stewart Brand.[33] In the context of academic computing as well, Joy Lisi Rankin has described the culture of academic time-sharing in the 1960s and 1970s as deeply social and collaborative, one in which computers were imagined as a communal utility.[34] Some privileged groups of users in the 1960s and early 1970s were able to experience passionate personal and social interactions with computing; but for most members of the public at the time, computing was more commonly associated with depersonalized and militaristic technology. For example, computer punch cards were incorporated into student protests in which they stood in for inhuman bureaucracies made more efficient by computer technology.[35]

It was not until the mid-1970s that personal computing became available as a commercial medium that could realistically be bought by individuals and used in private homes. This was helped along by the decreasing price and size of integrated circuits and the development of the microprocessor in 1971. Microprocessor technology—comprising chips that had the capacity for general purpose computing—was not developed with the application of personal computers in mind. But it nonetheless made it feasible to build small, relatively inexpensive, discrete microcomputers—a possibility that some hobbyists were to develop further.[36]

In the January 1975 issue of *Popular Electronics*, the Altair, a computer based on one of the newly affordable microprocessors, the Intel 8080, was advertised as a kit that consumers could purchase for under $400 and assemble at home.[37] The article accompanying the cover story announced, "The era of the computer in every home—a favorite topic among science-fiction writers—has arrived!" Of the possible applications for the Altair described in this issue, many were for commercial or industrial uses such as serving as an automated automobile test analyzer or navigation computer. Still, the article describes a few household uses: "sophisticated intrusion alarm system with multiple combination locks," digital clock with time zone conversion, and "automatic controller for heat, air conditioning, dehumidifying."[38] In reality, none of these applications was feasible for those purchasing this kit. For the $400 price advertised, the Altair did not come with the components we may now associate with a personal computer like a keyboard for inputs or a monitor for display.[39] The Altair is considered significant because it helped mobilize the energies and attentions of hobbyists, most famously the Homebrew Computer Club, where figures like Apple cofounders Steve Jobs and Steve Wozniak met. This informal group played a big role in focusing efforts on the development of the personal computing industry later in the decade. Yet, even as the announcement of the Altair described uses that potentially could be pursued at home, the computer was not yet promoted as part of family relations.

With the emergence of what some have called appliance computing, it became increasingly possible for users to own their own computer at home, even if they did not have skills to build it from a kit. In 1977 three companies released personal computers that could be purchased ready to use—the Apple II, the TRS-80, and the Commodore PET. Some computer magazines at the time like *Byte* took to referring to these as "appliance" computers to distinguish them from kits that required assembly by their users.[40] These appliance computers had a variety of origins: the Apple II was famously a product of hobbyists Jobs and Wozniak, while the others were produced by Tandy/RadioShack and the calculator manufacturer Commodore.

Once it was possible to buy a computer for the home, people wondered whether it had any practical uses in that setting. As early as 1979, commentators suggested that the hobbyist market had saturated—anyone who wanted to buy a computer simply for the excitement of tinkering or experimentation had already done so.[41] Would there be a market for home computers beyond hobbyists? In an attempt to answer that question in the affirmative, a list of common potential applications for home computing were repeated in dozens of popular articles, a few of which accounted for computing as part of family routines. Articles claimed that home computers could be used to inventory possessions like record collections, balance a checkbook, organize and file tax returns, print mailing labels, control lighting and heating in the home, plan a weekly menu, and store recipes.[42] Carl Helmers of *Byte* magazine optimistically claimed that although there were no useful home software applications yet, there soon would be.[43] Nonetheless, industry forecasters began to revise previously optimistic projections for the growth of the home computing market. Only two years after the Apple II first went on sale, a *New York Times* article opened with the question "What has happened to the home computer?"[44]

Throughout the 1970s and 1980s, skepticism about computers for the home did not fade completely, but reluctance to purchase computers was overcome to some degree as more software was developed and the computer industry continued to expand and lower prices. In late 1979, a startup called Software Arts released the spreadsheet software VisiCalc for the Apple II. Computerized spreadsheets were seen as a new and significantly powerful way to make use of personal computers. The program sold more than one hundred thousand copies by 1981 and soon other spreadsheet software products, like the popular Lotus 1-2-3, were also on the market.[45] The year 1979 also saw the release of WordStar. Although not the first piece of word processing software, it helped establish word processing as one of the advantageous applications of personal computing. (WordStar was surpassed in popularity by WordPerfect by the mid-1980s.)[46] In 1981, IBM, long associated with serious business computers and mainframes, entered the personal computing market. This, together with spreadsheet and word processing software, helped people see personal computers as powerful tools for work at home or for running a small business. As software and hardware developed to encourage consumers to invest in powerful, high-end personal computers for home offices, there was also rapid growth in the lower-priced end of the personal computing market. The semiconductor and microchip firm Texas Instruments (TI), the video game console pioneer Atari, and the toy company Coleco, among others, all introduced low-priced computers that might be suitable for homes and families.

Even as computers were sold in increasing number, this did not mean they were being more integrated with companionate family relationships or the feminized domestic space of the home. The higher-end machines maintained a masculinized aura associated with workplace or office uses that separated them from the everyday practices of family life. Attempting to distinguish themselves from the glut of new machines, representatives of companies producing higher-priced computers like Apple claimed that home consumers were not their market.[47] Popular home office software like VisiCalc maintained a masculinist association. The developer who created this spreadsheet application told a computer journalist that he imagined it like a video game and "saw himself blasting out financials, locking onto profit and loss numbers that would appear suspended in space before him."[48]

As home computing slowly spread, many newspapers and magazines continued to express doubts about the computer's suitability for domestic life. Many of the proposed uses for home computing did not immediately materialize or catch on with consumers. Critics pointed out that it was cheaper to use pen and paper or a pocket calculator to do some domestic tasks like balancing a budget or indexing recipes or record collections. Refuting the appeal of computers for home automation, one user reasoned that for $60 one could buy a thermostat that did much of the same work as a $600 computer to manage household appliances.[49]

Yet, some efforts were made to address consumers as part of a domestic space and family unit. The computer industry responded to the hesitance about home computing with multiple strategies to encourage purchases. In the early 1980s, stores like Sears, Macy's, Bloomingdale's, Kmart, and even Toys "R" Us started to sell home computers. Buying a home computer no longer required a trip to a specialty store; rather, it could be done at these outlets that would be more familiar to women. In the early 1980s, computer companies employed familiar celebrity spokesmen. Dick Cavett, Bill Cosby, Alan Alda, and William Shatner helped sell Apple, TI, Atari, and Commodore computers, respectively. These efforts helped situate computers as familial appliances by associating them with the stars of the already-feminized family medium of television. Sales of home computers in the early 1980s were also helped by further price cuts, partly caused by a "price war" waged between TI and Commodore.[50] Although this competition drove many companies out of the home computer market, it spurred a growth in sales overall. By October 1984, nearly seven million American homes (8.2 percent) were in possession of a home computer; of these households, 70 percent reported purchasing their devices in 1983 or 1984.[51]

Developments were also continuing in the software industry as publishers developed applications that were popular with users. Games like *Raster Blaster, King's Quest, Lode Runner, Castle Wolfenstein,* and *Microsoft Flight Simulator* were hits. Earlier in the decade, educators expressed concern about the poor offerings in the educational software market.[52] Over the course of the 1980s, many software firms developed successful educational titles (commercially, if not pedagogically) for the home computer, including *The Oregon Trail, Rocky's Boots, Number Munchers, Reader Rabbit,* and *Where in the World Is Carmen Sandiego?* Additionally, home office and finance software continued to sell well, including titles like Bank Street Writer, DB Master, and Personal Filing System. Home software proliferated and by the end of the decade approximately 15 percent of U.S. households owned a computer.[53]

At the same time, people worried that home computers would not be useful at home and would end up collecting dust, as suggested by the saying "a computer in every closet."[54] But this was not the only obstacle the computer industry faced. When home computing was avidly taken up in the home, some were concerned about how it would affect families. Women and girls were often at the center of anxieties about computing. The rise of home computers throughout the 1980s coincided with a widening of the gender gap in computer professions.[55] Lori Reed has discussed how the anxiety that women were being left behind was addressed in women's magazines like *Ladies' Home Journal* and through other means like Tupperware-style parties to teach women about computing.[56] Books like *Computer Confidence: A Woman's Guide* (published by Heller and Bower in 1983) sought to ease women's perceived fear of computing, and the Women's Computer Literacy Project offered seminars that taught women about computing on their own terms.

Women were also often described as collateral damage of men's computer addiction. As mentioned above, the computer widow was a ubiquitous figure in newspaper and magazine commentary that warned women about the dangers of computing in the home.[57] Women told of husbands distracted by home computers, and men reported the frustrated responses of their wives to their all-encompassing focus on the computer. Discourse about computer widows expressed the worry that interest in home computing would disrupt familial relationships. The fear was not only that computer users would be distracted from their spouses or children, but also that computer use would impair their ability to relate with their families at all. Psychologists worried that time spent with the impersonal and programmed communication style of computers would influence how users communicated at home; some warned of men "giving commands to family members as one does to the machine" or speaking to others using more "mechanical" communication styles.[58]

Resources were developed to help families manage the potential disruption to their relationships. *Family Computing* magazine, which debuted in September 1983, targeted an audience of readers who sought information and inspiration for integrating computing into family life.[59] In its monthly issues, the magazine documented real families who were finding a variety of everyday uses for computers that fostered their family bonds. A mother and daughter used their home computer to organize, research, and write a two-hundred-page family history; a woman used her computer to organize her own law school graduation party; a man used VisiCalc to help his mother decide whether to move, and his son used it to decide which prep school to attend.[60] Magazine contributors described using computers to entertain children for a birthday party or to write a family newsletter.[61] The magazine also offered simple programs users could key into their computers to create a digital family voting booth or graphic holiday display, make greeting cards, or monitor household phone use. Most of the programs described in these articles were not developed with domestic uses in mind. VisiCalc, for example, was business productivity software, but *Family Computing* showed that there might be ways to mobilize it to serve the companionate family.

In addition to these resources in magazines, software and hardware makers also provided programs and technologies to help accommodate computing to the family. Software companies created genealogy programs like Family Roots, Roots/M, and Your Family Tree. Computing was applied to home décor through programs like Home Improvement Planner. Other titles such as the Model Diet, the Eating Machine, and Nodvill Diet Program offered tools to plan family meals and manage different nutritional requirements. Psychological and self-help software, including titles like Understand Yourself, Mind Prober, Coping with Stress, and Timothy Leary's Mind Mirror, promised that computers could contribute to personal wellness. The Sears Home Control System was designed to allow computer users to automate their appliances. Although these domestic applications for computing are not all exactly like the examples of computing as medium of companionate relationality on which this book will focus, they are evidence of diffuse attempts in the computer industry to make sense of how computing could support companionate family life.

The Family That Computes Together

As mentioned above, *Family Computing* aimed to help families assimilate computing into everyday life. To advance this goal, in 1985 the magazine began running a contest looking for a Computing Family of the Year. Readers were encouraged to nominate their own or others' families and explain how they used computing together in "especially efficient, rewarding, or creative ways!"[62] The Mancinis, the first Computing Family of the Year, were inspired to buy an Atari 600XL after their son was exposed to personal computers at school. When installed at home, the computer first attracted the children who played games together—their mother, Kate, claims that in front of the computer was one of the few places the children did not argue. Next, Kate and her husband found uses for the computer in their consulting business. The article describes evolving uses for the computer: Through educational software, the whole family, including Kate's husband, was able to help the eldest daughter through her struggles with math. Kate used the home computer with a modem to complete her college degree remotely. The Mancinis' next goal was to invest in music software to enable them to spend leisure time together as a family in this creative pursuit. In her winning essay, Kate Mancini summarizes her feelings about family computing: "I feel that computers are fast becoming essential to any household . . . by enjoying the computer on a personal basis and together as a family we are helping our children to feel completely comfortable in this computerized world."[63]

The Mancinis, and many other families documented in *Family Computing*, demonstrate not only the effort made to assimilate computing into family life, but also how computers were seen in terms of larger challenges being faced by middle-class families in the 1980s to balance work, leisure, and companionship in the home. At a time when women with children were experiencing greater career pressures and entering the workforce in unprecedented numbers, the computer helped Kate Mancini advance toward her career goals at night while caring for her family during the day. In response to new expectations for fathers' intimate involvement with children, the computer allowed the Mancini patriarch to participate in his daughter's education. The computer also assured these parents that they were helping secure their children's future career success by providing them with computing skills, an urgent concern for families facing the reality of disappearing professional opportunities and a shrinking middle class. Furthermore, the computer was an entertainment device that the family could use to enjoy leisure time, bringing them closer together amid competing claims on their attention.

As depicted in *Family Computing*, the Mancini family represented a reconfiguration of a longer-standing companionate family ideal, one adapted to the cultural and economic context of the 1980s. This family is shown to be intimately connected, spending time together, sharing interests, and supporting one another while also balancing their varied individual commitments. But fulfilling the growing expectations for educational and career productivity in the home while maintaining strong affectionate bonds was not easy. Families needed help adjusting to new demands on their relationships, and computers offered one possible source of assistance.

The companionate ideal is not a natural or inevitable condition for family life. Rather, it is a historically specific formation that requires effort to sustain and is constantly subject to reformulation. The companionate family emerged as an ideal in the early twentieth century in the face of massive social and cultural transformations over the course of the previous decades. This included rising divorce rates, changing sexual mores, and transformations in traditional gender roles as women gained increasing access to public life, educational opportunities, and work outside the home. In response to these shifts that put pressure on existing family forms, the companionate family was presented as a structure better adapted to contemporary conditions of social life in the twentieth century. In the companionate arrangement, intimate bonds and more informal and expressive relations join family members. Spouses relate as friends and lovers and parents and children are friendly and affectionate. Marriage bonds are tasked with providing companionship and sexual fulfillment. Parental relationships with children are expected to be more democratic and nurturing of children's greater freedoms. This vision cast the family as a major source of interpersonal fulfillment for all concerned.[64]

In the 1970s and 1980s when computing was first becoming available for home use, the companionate family ideal was being reformulated in response to new challenges. Historians have discussed the late 1960s and succeeding decades as a time of transformation in the American middle-class family, promulgated by economic transformations and social movements like women's liberation, gay liberation, and Black civil rights.[65] Family demographics and kinship structures were being affected by the shift to a postindustrial information economy. Rising inflation, high unemployment, and loss of the family wage made a dual income increasingly necessary for the maintenance of a middle-class family lifestyle.[66] The demographic changes that began to be noted in earlier decades continued into the 1980s. Reagan-era economic policies led to a shrinking of the middle class and contributed to the uptrend in dual-income families. By the late 1980s, such families made up 58 percent of married couples with children.[67]

One significant result of these shifts was the rising number of women with children working outside the home. By 1986, 54 percent of women with children under six years old were in the paid labor force, compared to 31 percent in 1970 and 12 percent in 1950.[68] This change in gendered labor patterns had wide-ranging repercussions for companionate family arrangements. With more women working outside the home, the 1980s was marked by a "daycare crisis" in which single-parent and two-job families struggled to find support for childcare.[69] Studying two-job families, Arlie Hochschild drew attention to what she called a "stalled revolution" occurring in that decade. Even as women with young children were increasingly entering the paid labor force, workplaces and homes were not adapting to this transformation. These middle-class working women were being tasked with a "second shift" of homework on top of their paid labor outside the home.[70]

Adding to increased expectations on women to labor both inside and outside the home, the 1970s and 1980s saw the emergence of "supermoms" discourse. This phenomenon was only one aspect of what Susan Faludi has referred to as a backlash in response to feminist gains.[71] Women, who might have been pursuing professional careers, were bumping up against an intensified ideal to embody a perfect and zealous motherhood. This included new child-rearing values that required mothers to schedule and engineer every aspect of their child's time for maximal productivity and enrichment. In other words, supermoms were expected to help raise superbabies—kids who were speaking French, studying violin, and being drilled in mathematics all before they learned to read.[72]

Amid this increasing attention on childhood achievement, fathers were a potential source of childcare; however, as Hochschild found, they rarely took on substantial responsibility for this work in practice. This period saw shifting ideals for fatherhood emerging, including what E. Anthony Rotundo describes as a new ideal of participant fatherhood. This version of fatherhood mandated "active and engaged participation of a man in all facets of his children's lives" as well as "intensive emotional involvement."[73] Changing expectations for men's contribution to family life were also developing in response to the demands of feminist activism. Men were navigating changing expectations to develop affectionate relationships in the family and to take on some household duties.[74]

The makeup of familial structures and intimate bonds were changing in other ways, too. Expectations for sexual and intimate relationships for couples and spouses had been significantly transformed after years of feminist contestation and sexual revolution. Marriage and romantic relationships in the 1980s were being tasked with supplying fulfilling sexual and emotional experiences for both members of the couple. Partners were charged with working on communication and intimacy while also maintaining sexual and romantic passion.[75] They could also find easy access to adult videos and sexual aids such as vibrators, sex manuals, and couples' porn.[76]

The computing and electronics industries were among many that responded to these societal shifts. James Hay has argued that new technologies in the 1980s helped reconfigure the home as a "(neo-)liberalized" domestic sphere—a space of autonomous self-governance—and made way for the later adoption of personal computers. The influx of technologies like smart TVs, remote controls, VCRs, and programmable coffee makers, dishwashers, stoves, fitness systems, and alarm clocks promised that "their users could rely on these devices to achieve both greater flexibility and greater management of the household and everyday life. These were machines that could be seen as self-governing, for households that were self-governing. The domestic sphere itself became a mechanism that freed their users to do other tasks."[77] Hay analyzes how digital technology fostered fantasies of efficient households and provided tools for neoliberal self-governance. However, computing in the 1980s also offered tools to bolster familial bonds and to aid families and individuals as they navigated changing expectations for intimacy and companionship. The computer industry addressed products to the peculiar needs and challenges faced by the middle-class family of the 1980s, including anxieties about changing norms for masculinity, increased expectations for coupled relationships, and changes in childcare and child-rearing. Computing technology was seen as a potential resource for families struggling to manage their relationships with one another as they navigated economic and social demands and individual desires. Rather than disruptions or challenges to family life, these applications of companionate computing were offered as tools to help support the white, heteronormative, middle-class family structure.

Chapter Outline

The above discussion shows that the companionate family ideal was subject to major social and cultural shifts, and computer applications were responding to these shifts in turn. To address the breadth of what it means to say that computers were imagined as media for domestic relationality in the 1980s, each chapter is centered on a different type of companionate relationship and the associated ways that computing was imagined as a medium to modulate such relationships. This includes computer as mediator of the couple, computer as support and extension of participant fatherhood, computer as proxy for maternal child-rearing, and computer as critic of an instrumental form of heteromasculine seduction. *The Intimate Life of Computers* starts with a chapter on applications designed for two simultaneous users and addressed most explicitly as an intervention into a companionate relationship. As the chapters progress, my analysis seeks to demonstrate how programs and products that do not immediately appear to be involved in companionate relationality were often designed and promoted as technologies that could in fact mediate and supplement familial and intimate relationships.

The first chapter, "A Ménage à Trois with Your Computer: Romance Software as Mediator for Couples," discusses therapeutic programs made for couples to work on their romantic relationships. As alluded to above, couples in the 1980s were struggling with increased burdens on their time and attention, as dual-career families became a more common phenomenon. Historians have also noted that during this period relationships were tasked with greater expectations to provide intimacy and sexual fulfillment.[78] Computers threatened to add to these challenges by distracting men from their wives and families. In response to this, some software companies developed programs that could work as solutions to enhance rather than detract from couples' romance and intimacy. Through close readings of this romance software, like *Interlude* (1980), *Lovers or Strangers* (1982), and *IntraCourse* (1986), I show how these programs guided couples in sharing and processing their potentially conflicting romantic and sexual preferences and desires. In these programs, the computer acted as a mediator by probing for information about the couple's relationship and coordinating how it would be distributed and operationalized. These programs drew on feminized therapeutic and self-help culture and also reflected changing norms of companionate intimacy and sexuality. The developers marketed this software to couples, not isolated men, and imagined how the potentially individualizing or customizing qualities of computers could be used to strengthen relationships rather than to alienate users further from each other.

The availability of cheaper microprocessors in the 1970s and 1980s boosted the market for microcomputers and associated software. Microprocessors also proliferated in homes embedded into a variety of other embodiments. The second chapter, "'Not an Appliance, but a Friend': Personal Robots and Participant Fatherhood," discusses one such application of microprocessor technology—personal robots for the home, often described as personal computers on wheels. While romance software was obviously designed for multiple simultaneous users, personal robots appear at first glance to have been designed for individuals to operate on their own. Nonetheless, personal robots were framed as a shared family computing tool. Robot promotions were primarily addressed to men, but showed them how they could use robots to enrich family life. Computing was pitched in the 1980s as a technology for professional development that could be pursued privately from home. However, as Elizabeth Patton notes, this didn't necessarily mean that computers integrated with family routines.[79] Part of the appeal of personal robots was that they could offer private training in robotics and programming for men at home. Although appealing to men might further alienate them from domestic life by confining them to workshops or dens, personal robots were offered as tools to help men share their interest in computing with their families, especially their children. This was appealing at a time when expectations for men's participation in child-rearing and domestic life were changing. This chapter discusses the marketing and design of personal robots by Heath, Androbot, and Hubotics, to show how personal robots served as technologies of domestic relationality that could support and extend their users as they tried to achieve the ideals of participant fatherhood. Furthermore, the way these robots were styled and promoted shows that robot makers foresaw their potential impact on women and family life. There was clearly an awareness that users would need to navigate their robotics hobby within the constraints of their domestic relationships and responsibilities.

The following chapter, "'A Doll That Understands You': Computer Talking Dolls as Parenting Proxies," continues the discussion of microprocessor-embedded products, exploring the incorporation of these technologies into talking and listening toy dolls. Whereas robots were offered as computing technologies that supported or aided fathers in extending care and attention to family, computer talking dolls can be better understood as stand-ins or proxies for maternal care. Dolls were designed for girls to play with, but their promotion also targeted parents, especially mothers. It spoke to their perceived desires and anxieties relating to child-rearing, which were made worse by contemporaneous discourses about the burdens on supermoms and early childhood achievement in which computers were often invoked. Computer talking dolls promised to serve as a proxy for maternal attention by simulating an autonomous awareness of the girls playing with them—a simulation enabled by digital speech- and sound-recognition technologies. Designers drew on traditions of doll play to imagine a form of computerized play that was more socially embedded in the home, especially when compared to video games, another form of computer play that was more popular at the time.

Chapter 4, "Sex and the Singles Game: Adult Games, Cringe, and Critiques of Masculine Seduction," returns to a discussion of desktop computing in the form of adult games that simulate seduction and are designed ostensibly for single players. As single-player games about seducing digital women, adult games may initially seem to be the furthest from the view of computing as a medium of domestic relationality; instead, they may be seen as representing the kind of isolated, masculine attachment to computers that might replace or impede rather than augment social relationships. Yet, even these games can be interpreted as part of a framework in which computers function as technologies to mediate companionate relationships. Focusing on two adult games, *Leisure Suit Larry in the Land of the Lounge Lizards* (1987) and *Romantic Encounters at the Dome* (1987), this chapter argues that beyond simulating heterosexual seduction, these games also critique a certain instrumental style of heterosexual masculine relationship with women. Rather than present their player characters as ego ideals, *Leisure Suit Larry* and *Romantic Encounters* instead rely on cringe aesthetics that promote reflections on masculine norms and modes of relating with women. Although not necessarily representing radical critiques of masculinity, the approach these games take to denaturalizing masculine representation and instrumental gameplay demonstrates the influence of feminist critiques on the development of games. Although these games are not presented as feminist correctives to make men more sensitive or even pick-up guides to help them get better at seduction, they nonetheless offer an instigation to reflect on norms of masculine heterosexuality.

Together these chapters illustrate a variety of attempts made to use computers to mediate relationships and buttress the companionate family ideal as it was being redefined in the 1980s. But as the discussions in these chapters will show, these attempts were marked by ambivalences. Tensions arose both due to the intense cultural pressures on the companionate family at the time and the computer's dual role in addressing users individually and as part of companionate relationships. The efforts to mobilize computers as companionate technologies was not always successful or definitive, but is rather ongoing. Pointing forward, the coda discusses some of the ways ongoing contestations over companionate relationships and computer interactions can be seen in more contemporary examples of digital media applications.

Although it is now commonplace to think of digital media integrated into domestic life, the computer applications developed in the 1980s represent an early attempt to envision computers as tools for companionate relationality at a time when meanings attributed to family were changing. Not all these applications were commercially successful, but they show how computer culture was attempting to respond to contestations over the meaning of family life and what role computers would play in that structure.

1 A MÉNAGE À TROIS WITH YOUR COMPUTER

Romance Software as Mediator for Couples

RELEASED IN 1982 FOR THE APPLE II, *Lovers or Strangers* offered couples the means to turn their home computer into a romantic device. Advertisements for the program appeared in computer magazines such as *Popular Computing, Softalk,* and *Creative Computing.* However, they deviated from the imagery of action-adventure games or productivity and home office software more commonly circulated in these publications. Instead, the advertising for *Lovers or Strangers* resembled the aesthetics of popular romance novel covers: it featured a pair of lovers before a dreamy sunset whose embrace is threatened by a prominent tear running down the center of the image (Figure 2).

To play *Lovers or Strangers,* two users are invited to sit side by side in front of a desktop computer, sharing a keyboard between them. Questions covering topics related to sex, religion, and work appear on the screen one at a time for the couple to answer. Reminiscent of television dating and newlywed games, players are asked to anticipate their partner's response in addition to volunteering their own. Subsequent questions appear on screen only when both partners have answered for each other and for themselves. During the interview, the couple does not receive immediate feedback from the computer about how accurately they have judged their partner or how aligned their answers are with this person sitting beside them. Instead, they are held in suspense as they respond, the pace of their progress dependent on each other's readiness to reply. At the end of the interview, the program processes their responses, scores their compatibility, and delivers advice for how they can learn more about each other. Yet throughout the program the computer has already encouraged the couple to spend time together devoted to considering their relationship, and the software manages and coordinates the users as they do so.

Figure 2. An advertisement for *Lovers or Strangers* resembles the aesthetics of romance novel covers. The spot was published in the January 1983 issue of *Creative Computing*.

Lovers or Strangers belongs to a genre of programs developed for personal computers in the 1980s that I am calling "romance software." These programs drew on contemporary psychological and therapeutic discourses about sex to offer couples advice and suggestions for more satisfying relationships. They also managed users' interactions with the computer as a romantic activity. At a time when some commentators worried that computers were distracting men from their social obligations and creating computer widows, romance software represented the possibility that couples could use the computer together to enhance their relationships.

This chapter considers how romance software in the 1980s marshalled the home computer as a tool that could help mediate relational and communication ideals and bolster the heterosexual couple. To do so, I situate these programs in the context of contemporaneous discourses that were redefining coupled ideals for intimacy, romance, and sexual relating in marriage in response to feminist critiques, economic changes, and other cultural pressures.

Analyzing three distinctive programs—*Interlude* (Syntonic, 1980); *Lovers or Strangers* (Alpine Software, 1982), reissued as *Friends or Lovers* (Softsmith, 1984); and *IntraCourse* (Intracorp, 1986)—this chapter shows how romance software attempted to mobilize computing to help couples negotiate competing demands, to express and resolve individual desires with coupled harmony, and to balance the imperatives to disclosure necessitated by intimacy discourses with the unpredictability and excitement required to maintain romance and passion. They did this not only by offering advice ostensibly optimized by computational means, but by presenting the computer as an active mediator in the couple's experience. In these programs the computer selectively probed for and distributed information and helped organize the interaction between paired users. Romance software largely addressed users as part of a heterosexual, monogamous couple. But as this chapter discusses, even as these programs attempted to manage conflicting desires and utilize the computer to strengthen coupled bonds, they also risked reinforcing normative gender relations or suggesting more promiscuous forms of relation beyond the couple.

Romance software was presented as a way for couples to bring their sex lives closer to ideals of the period. To reach an audience of couples, rather than isolated male users, these programs referenced contemporary therapeutic and self-help discourses and software makers relied on associations with women's culture in their marketing and development.

Even if the resulting software did not primarily appeal to a large audience of women or successfully center their concerns, the influence of women's culture and feminist critiques of the couple still shaped the way these programs were designed and circulated. Most prominently, this influence can be seen in the way these programs situate computing as a medium for negotiating intimacy and sexuality, work that has long been associated with women.

Scholars and historians often depict personal computing in the 1980s as atomizing, representing an abstracted masculine individualism.[1] This can be seen, perhaps most prominently, in the widely circulating discussions of computer widows—depicted as women who were being neglected and had to work to sustain their husband's interest in competition with computers. The existence of romance software demonstrates that some software makers were imagining personal computers not as individualizing or atomizing, but rather as a technology that could enhance domestic relationships and sustain the companionate family by working on coupled romance and intimacy. These programs addressed their users not as individuals, but as couples seeking romantic experiences and offered them a fantasy of heterosexual romance and sex enabled and intensified by the computer.

Managing Romance and Sex in the 1980s Couple

The introduction of computers into middle-class domestic spaces in the 1980s inspired widespread discussions about the phenomenon of computer widows. Circulated in newspapers and magazines, accounts of computer widows described women whose husbands had become so preoccupied by their computer hobby that they neglected their families, and especially their wives—even losing interest in sex with their partners.[2] Most often, the computer widow was a specter invoked in articles describing potential social effects of computing. But occasionally these women were given the opportunity to speak about their experiences with computers in the home. In an issue of *Family Computing* magazine, one such computer widow, J. D. Sidley, tells a story of her marriage transformed by computing. Once, her husband used to give her lovingly handwritten sonnets, now he passes her computer-printed grocery lists and spends so much time using the spreadsheet program VisiCalc that he doesn't notice her negligee-clad attempts at seduction. In her account, Sidley diagnoses the problem with computers as their cold practicality and lack of romance.

As she continues, though, the story of marital alienation gets more complicated. Sidley explains that she and her husband used to each come home from work and communicate about their respective days on the job. But now she has "taken a detour through Babyland, U.S.A." Difficulties in communication seem to be as much about career and childcare and their impacts on marriage as they are about the distraction of the computer. Sidley even admits, "You tell me that this silent machine isn't to blame for the growing rift in our relationship? Well, I'll tell you, it ain't helping, sister."[3] Throughout her account, Sidley allows resentment and even humorous anger about gender roles and marital relationships to leak through. However, like many accounts penned by computer widows, she ends on an optimistic note. Her husband used a computer to type out a love sonnet, and Sidley concludes that all that computers—and marriages—need is a little romance.

As this computer widow's account suggests, computers in the home were one factor in larger changes that shifted the way couples related to each other and managed labor, romance, and sex. At this time, relationships between couples and spouses were changing in fundamental ways, and what it meant to be part of a couple was being renegotiated. The 1980s saw an increase in dual-career couples, a high but stabilizing divorce rate, and a popular culture offering more sexualized fare.[4] An increasing percentage of women working outside the home motivated reconsiderations of gendered divisions of labor, as did feminist contestations over labor and sex roles.[5] At this time, too, transformations in popular and sexological conceptions of sexuality were increasingly incorporated into expectations for marital sexual relations.[6]

Romance software was only one example of a proliferation of products and services available for couples looking to enhance their sexual relationship and manage raised expectations for sexual fulfillment. Over the course of the 1970s and into the 1980s, popular culture had become increasingly sexualized. As Elana Levine has discussed, the proliferation of sexual culture and "commodified sexual expression" had become so widespread that it even found its way into the family medium of television.[7] Couples navigating their sex lives had access to a growing range of resources to develop their sexuality in the form of sex manuals, toys, and lingerie. Jane Juffer describes how vibrators, although they had been available long before this period, gained increasingly mainstream status as aides to women's sexual fulfillment. This was part of a larger legitimation and popularization of discourses about women's masturbation and orgasm, including the publication of texts like Nancy Friday's 1973 compilation *My Secret Garden*.[8] Juffer also describes the proliferation of other examples of what she calls "domesticated porn," including self-help books and couples' sex videos, that were part of an effort to speak to women's sexual desires.

Major transformations in sexual relating had developed over the course of the 1960s and 1970s, especially with respect to understandings of women's sexuality. These changes increased the pressure on marriage and couples to provide sexual fulfillment. In their study of sexual culture after the sexual revolution Barbara Ehrenreich, Elizabeth Hess, and Gloria Jacobs describe how authors of sex manuals and other experts "began to redefine sex as a variety of options from which the savvy couple could pick and choose."[9] Popular sex manuals like *The Joy of Sex* (Alex Comfort, 1972) and *The Sensuous Woman* (Terry Garrity, 1969) depicted sex as consisting of an assortment of acts rather than prescribing a single ideal of heterosexual activity. Sex was seen as a practice or set of techniques that could be developed and continually improved on, with the help of consumer goods. By the early 1980s, these views about sexuality for heterosexual couples had entered the mainstream. Ehrenreich, Hess, and Jacobs even describe Tupperware-style parties for married suburban women that offered light S/M gear and sex toys for sale in the comfort of their own homes.[10]

Connected to shifts in sexual relations, communication and intimacy were also intense topics of interest and debate for couples in the 1980s. David Shumway argues that the last quarter of the twentieth century saw the emergence of what he refers to as intimacy discourses as the dominant way to make sense of marital relationships and heterosexual coupling. Developed within expert and self-help literature as well as in films and novels, intimacy discourses placed intense pressure on couples to work on their relationships through expressive sharing and communication. The focus on intimacy was a relatively new way to understand relationships, and Shumway argues that it superseded romance as the way to judge the strength of coupled bonds: "Romance offers adventure, intense emotion, and the possibility of finding the perfect mate. Intimacy promises deep communication, friendship, and sharing that will last beyond the passion of new love."[11] Making it even harder on couples, Shumway notes that romance and intimacy coexisted as demands on the relationship. As they were expected to share fully their deepest thoughts and desires, couples also needed to somehow maintain some mystery and spontaneity. Experts, relationship guides, and self-help literature told couples to communicate more effectively but also to labor to maintain romantic and sexual passion amid this increasing closeness and intimacy.[12] Romance software was one tool available to assist with these demands.

These renegotiations of intimacy and sexuality within marriage and coupling were taking place at the same time that economic pressures were also changing labor patterns within the family. One of the major changes to family life was the increase in numbers of women working outside the home and the rise of dual-career couples. By the mid-1980s, two-thirds of all women with children were in the workforce. The rise of dual-career couples meant that middle-class families were experiencing greater burdens on their time. Arlie Hochschild documents the way this "speed up" in family life in the 1980s placed a disproportionate burden on women, who had to carry out a "second shift" of homework in addition to their jobs outside the home.[13]

Part of the second shift that continued to be borne primarily by women was work on maintaining marital and romantic relationships. Kristin Celello has described a long history in which cultural experts figured marriage as "work," with the burden of working on the marriage falling primarily on women. Even though women had been in the workforce long before the 1980s, the increasing cultural prominence of dual-career couples led to renewed anxieties about the need to work on marriage. As Celello argues, one of the concerns circulated in response to women joining the workforce was that "between their paid and unpaid labor, wives had little time to concentrate on having pleasurable sexual relationships with their husbands." Women's magazines and advice books often singled out dual-career couples as most at risk; however, they suggested these types of marriages could be successful if couples, and especially women in these couples, were willing to work to sustain the relationship.[14] This included effort to meet increasing demands on marriage as a source of sexual fulfillment. The many accounts of computer widows spending time learning computers to save their marriage were typical of this pattern. This suggests, too, that even if romance software was targeted to men in some instances, it still related to the labor of supporting the couple, a task often viewed as women's work.

Overall, these cultural and economic changes created intense and sometimes competing demands on couples. Partners should share everything and be expressive but still maintain some mystery and romance—all while pursuing greater expectations for vigorous sex lives. Within the couple, partners were encouraged to pursue their individual desires for fulfillment without sacrificing their commitment to each other and to the coupled unit. And, particularly for women, they needed to do all this work on their relationships while also balancing job demands.

Anthony Giddens has argued that this period saw the emergence of what he calls the "pure relationship," which is one "entered into for its own sake, for what can be derived by each person from a sustained association with another; and which is continued only insofar as it is thought by both parties to deliver enough satisfactions for each individual to stay within it."[15] Related to this, it was not unusual to be a part of more than one sustained romantic and sexual relationship in a lifetime. By the 1980s, it was more common than in previous decades for heterosexual couples to cohabit with a partner to whom they were not married: the number of unmarried couples living together in the 1980s had quadrupled since 1960.[16] The divorce rate also remained high throughout the decade. Individuals unhappy within their relationships or marriages could divorce and seek other ways to satisfy their romantic and sexual desires. Only an emerging ideal in the 1980s, most relationships did not fully adhere to this model. Yet, many cultural discourses converged to place increasing demands on the couple and to allow for individuals to seek out different routes to meet these expectations.

The introduction of computers into middle-class domestic spaces in the 1980s had the potential to exacerbate the growing demands on the time and energy of married couples and pose further obstacles to marital harmony and sexual fulfillment. But computers could also be viewed as tools to overcome these same challenges. This is apparent, too, in the popular discussions about computer widows mentioned above. In their accounts, computer widows addressed problems related to communication, incompatibilities between career and domestic interests, and sexual intimacy, all of which they felt were affected by the entrance of computers into the home. In suggestive ways, computer widows gestured to larger cultural pressures on marriage.

For example, Robin Raskin, sharing her story with *Family Computing* magazine, describes how, at first, computers only made things harder for her and her husband. Raskin and her partner were living on a houseboat; she was busy raising a young child and pregnant with a second when her computer scientist husband brought home a computer and placed it "next to our bed in the one bit of space we had on the whole boat!"[17] Like

Sidley, she communicates both resentment and fear as conversations with her husband dried up and their interests diverged. But also like Sidley, it appears this is not only due to the presence of computers, but characteristic of changing expectations for gender roles and marital intimacy in society at large.

Although the entrance of personal computers into the home was one factor that may have exacerbated the time pressures on the couple and potentially disrupted their efforts to develop greater intimacy, the computer was also posed as a solution to these problems. In fact, this is exactly how Raskin concludes her account, with advice for women to learn to share their husbands' interest in computing to save their marriage or improve their relationship. Raskin's solution was not to try to regulate her husband's computer use, but rather to learn about computers herself: "I knew marriage and family was important to me, and I knew I was going to work at it. I knew the computer was going to be our link. . . . So I learned about computers to save our marriage."[18] This reformed computer widow goes on to advise other women to take initiative in asking husbands to describe their work on the computer or to try to use software or play games that their husbands enjoyed.

Raskin was not alone in expressing a belief that the technology could bring couples together. Another woman explained how the excitement of learning computers with her husband helped keep the "fires burning" in their relationship.[19] Other stories circulating in computer magazines tell of spouses who were brought closer together in pursuit of a computing business that they ran together.[20] As all these examples demonstrate, the solution proposed to the potential distraction caused by computing was to turn it into a couples activity and bring romance to the computer.

With these prominent cultural anxieties circulating about computers as potential threats to couples, software developers were poised to respond with consumer products to enable couples to use the computer to strengthen rather than disrupt their intimate and romantic bonds. Romance software provided the means through which couples could turn computing into a joint activity. Rather than simply advising women to overcome their computer phobia or indifference, these programs fostered a coupled experience around computing that could be purchased ready to use. Programs like *Interlude, Lovers or Strangers,* and *IntraCourse* not only brought couples together but mobilized the computer to help make heterosexual relationships stronger and more exciting, promising to engage the couple in an erotic encounter with their desktop.

"A Cheap Rip-Off or an Entertaining Product"

The lament of computer widows about the invasion of computers into their homes shows that the devices had already made their way into many bedrooms. Yet, when some software companies responded with programs that made the connections between computers and domestic sexuality explicit, this was not always welcomed. Commentary in general interest newspapers and specialist computer magazines questioned how appropriate it was to apply computers to the area of sexual relationships; sometimes these products were dismissed as ridiculous jokes. As one *Chicago Tribune* reporter observed, "It was probably inevitable that the computer age would someday mesh with the sexual revolution. But now that it's happened, it doesn't seem any less weird."[21]

Software for sex and romance was often discussed derisively, but *Interlude,* released in 1980 when the market for home computers was still quite young, was one of the programs that seemed to generate the most ridicule. This was not helped by the sensational advertising that publisher Syntonic used to market the program. An advertisement for *Interlude* that circulated widely in computer magazines featured a woman in lingerie sharing her bed with a microcomputer.[22] In the image, the woman and the computer are visually equated as sexualized objects—both lie atop satin sheets and are turned toward the viewer.

In 1980, many in the personal computing industry still expected that computers' target audience consisted largely of male hobbyists. It was not until a boom in sales in 1983 and 1984 that home computers reached even 8 percent of U.S. households.[23] When discussing *Interlude,* articles often referenced the program's sensational and sexualized advertising as its distinctive characteristic and evidence that it was targeted to the prurient interests of single heterosexual men.[24] A reviewer for *Interlude,* after carefully describing the ad and making the requisite sex joke, asserts, "Everyone who's seen the ad must want to know what *Interlude* is: a cheap rip-off or an entertaining product."[25]

If all examples of romantic software were just a "cheap rip-off" to get the interest of juvenile or sex-starved computer users, then it might be tempting to dismiss them as a sidenote in computer history. In fact, existing scholarly histories of computing and gaming rarely address examples of romance software. When programs like *Interlude, IntraCourse,* or *Lovers or Strangers* are discussed at all, it is primarily as examples in popular internet lists in which they represent misguided or ridiculous notions about computing in the 1980s.[26] In other words, these programs are often treated as a joke now, just as they sometimes were in the 1980s.

It is not difficult to see why these games might be overlooked. Aesthetically, they are often rudimentary, consisting mostly of unadorned text on monochrome backgrounds (Figure 3). Visually the examples of romance software discussed in this chapter are largely indistinguishable, all featuring questionnaires administered in the form of menu-based screens from which users select their responses. The programs lack complex moving images and, unlike 1980s text-based adventure games that have inspired popular nostalgia and scholarly attention, romance software does not provide complex narratives or vivid descriptions of spaces to explore. Furthermore, other than *Interlude*, it does not appear that these programs were commercially successful.

Yet dismissing romance software for its sensational advertising, sparse graphic style, or poor sales record risks ignoring the compelling ways these programs imagined computing as an activity that mediated sex and romance by appealing to romantic couples. Furthermore, disregarding these programs as simply the puerile products of a masculine computer industry overlooks how they were affected by contemporaneous contestations over companionate relationality. Sexual and romantic ideals for the heterosexual couple addressed by romance software had been

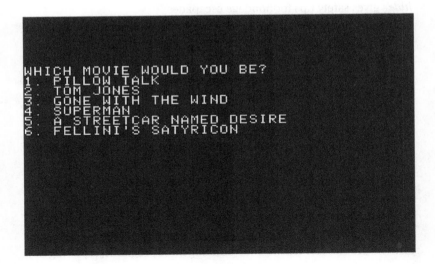

Figure 3. Screenshot from an emulation of *Interlude*. The program uses text-only menu screens.

profoundly shaped by feminism and other social forces. These impacts can be found in the resulting software that seeks to contend with these relationships, though often indirectly through their incorporation of popular and therapeutic sex discourses. Even if these programs were ridiculous at times, they also represent examples in which contemporaneous discourses about the family and sexuality were being worked through by the software industry.

Moreover, the reason romance software like *Interlude, Lovers or Strangers,* or *IntraCourse* would be considered a joke (both now and when they were released) is not simply because they advertised themselves as fulfilling men's sexual desires. There is a tension apparent in discourses from the 1980s that critique romance software. For example, one software critic writing for the computer magazine *inCider* offers a critical review of the sequel to *Interlude* that chastises the program for its "crass wisecracks" and "sniggering" tone. Even as he claims that his disapproval of *Interlude II* (1986) is due to its immaturity or lack of seriousness, he notes of the advice offered by the program, "A few are as mild as 'Spice Up Your Marriage' hints from *Redbook;* others would fit into *Cosmopolitan;* and the rest— well, one of *InCider*'s least feminist employees (he goes to strip clubs on road trips) could only cry, 'Gross!' It's nice to take a break from software manuals that discuss back-up and hardware requirements, but *Interlude*'s gives safety tips for bondage escapades."[27]

For this reviewer, both strip clubs and bondage seem to cross the line of appropriate sexuality, but one wonders what it is that his "least feminist" coworker finds so gross. Could it be the feminized mode of sex advice that made it feel like he was peering into the pages of women's magazines? Discussions like these raise questions about the inspiration for critiques of romance software. Is romance software ridiculous because these programs are the attempts of juvenile or unenlightened men to use the computer for sex—like the sexually suggestive *Interlude* ad might indicate? Or are they ridiculous because of their association with the long-denigrated imagery and thematics of women's culture that attempt to allow for the expression of women's sexual desires—featuring sex advice that would fit within the pages of women's magazines or, in the case of *Lovers and Strangers,* advertised with imagery clearly resembling romance novel covers? The debates and tensions surrounding these programs, exploring how to use computers to define and encourage good relationships and good sex, suggest that further investigation is needed.

Romance Software's Relational Ideals

Even for those software companies pursuing the idea of using computers to enhance romantic relationships, it was not obvious how computers could best help and toward what ideals of coupled relationality they should strive. All three programs I will discuss show the influence of contemporary shifts in discourses on intimacy, romance, and sexuality that had been shaped by histories of feminist struggle over gender roles and companionate relations. Yet, as these programs mobilized the computer as a mediator that could improve coupled relationships, they conceived of different presumed ideals and imagined the computer's role differently. *Interlude* and *IntraCourse* were focused primarily on augmenting sexual aspects of romantic relationships, whereas *Lovers or Strangers* included questions about sex as one aspect of relationship compatibility.

The programs differed not only in their thematic focus and advice but more significantly in the ideal versions of heterosexual coupling they sought to encourage for couples and how they imagined the computer as a mediator to achieve these goals. *Interlude* made small gestures to ideas about the need for improved communication, but it leaned much more heavily toward presenting romance, play, and novelty in sex as the key to better relationships and offered the computer as an instigator of sexual play. *Lovers or Strangers*, in contrast, valued confessional or self-expressive discourses in which the computer's main function was to encourage conversation between the pair to increase their intimacy with each other. *IntraCourse* presented the computer as a data processor and information service that would help users unlock hidden truths about individual sexuality and the interpersonal techniques necessary for a fulfilling sex life. Contemporaneous contestations over fulfilling heterosexual relationships came to be refracted differently in these programs, but they were all shaped by these shifting discourses about romance and sex.

When it appeared in 1980 *Interlude*, released for the Apple II and TRS-80, was likely the first example of romance software for couples developed for home computers. *Interlude* was sometimes critiqued for its seemingly unserious or snickering approach to sexuality, but in fact, the program leaned into this attitude as part of its therapeutic value. The *Interlude* manual opens by declaring that "sex is *adult's play!* [sic] It is an area in which we as adults can release our inhibitions and just *play*. We can indulge our fantasies, explore new sensations, play games with each other—all within the private boundaries of an intimate relationship."[28]

Even as it made occasional references to ideas about intimacy and communication as necessary for a healthy coupled relationship, the program was most focused on cultivating sexual novelty, variety, and play. As sexual activities were structured by this program, communication between the couple was not always prioritized; surprises and novelty were frequently given more emphasis. This ideally could benefit the couple by helping them fulfill contemporary imperatives to maintain greater romantic passion and variety in their sex lives. But it also risked using the computer to reinforce unequal gendered power relationships within the couple by investing greater agency in the decision making of a computer program developed within a masculine software culture.

Interlude consisted of an onscreen questionnaire that asked about the sexual desires and preferences of users to offer tailored recommendations for sexual activities for the couple to try. Although Syntonic's advertising, featuring sexually suggestive imagery of a woman in bed with a computer, suggested that it was addressed primarily to heterosexual men, the program itself was structured for a heterosexual couple. *Interlude* was designed to accommodate two users interacting with the computer together in a joint session. The program assumed heterosexual couples only, it could not accommodate more than two users, and it asked at the start for the name of one male player and one female player to whom it addressed its questions. *Interlude* could also be used by a single user, but the resulting recommendations were still designed for a heterosexual couple. Based on the answers provided to the computer interview, the program then directed users to one of over a hundred "interludes" printed in an accompanying manual. Interludes were one- or two-paragraph descriptions of sexual activities for the couple to try.[29]

Interlude promised to use computational methods to direct couples to optimized sexual relations based on their answers to onscreen questionnaires. The program poses questions that aim to assess the level of sexual desire and the preferences of each member of the pair. For example, the interview asks users their sexual "attitude" on a spectrum from "I'd rather watch TV" to "Take me, I'm yours!" Some questions are posed more indirectly, such as one that asks users to compare their sexual identity to a cat, ranging from "pussy cat" to "tiger," or to a movie, ranging from *Pillow Talk* (dir. Michael Gordon, United States, 1959) to *Satyricon* (dir. Federico Fellini, Italy, 1969). The program also asks specific questions about items or activities to include or exclude from the interlude—such as special clothes, light sadomasochism (S/M), or dirty books or movies—

and queries the users' interest in different erogenous zones. Answers from the questionnaire appear to influence which interlude the program chooses for the couple, to the degree that the intensity or complexity of interludes would be tailored to the level of sexual interest the users divulged. Additionally, certain types of activities or accessories should only be included in the suggested interlude if users did not protest to their inclusion when asked.

For example, if both users answer that they are not interested in sex or ask for interludes that are short and simple with no surprises, this will exclude the more sexually adventurous interludes. The program might instead suggest "Win Some, Lose Some," which directs the woman to curl up with a good book and leave sex for another night; or "The Eyes Have It," which instructs the man to pretend he is alone and look at pornography and masturbate in proximity to his partner, who might presumably enjoy watching him. On the other hand, if both partners answer the questionnaire indicating that they are very interested in sex, describe themselves as "tigers," and report a desire for a complex interlude or one involving "special clothes" or "mirrors," the program may direct them to the interlude called "Champagne Hour." This interlude is directed to the male user and suggests that he prepare champagne and a bubble bath to share with his partner before having sex with her bent over the bathroom vanity and watching herself in the mirror.

The published interludes are typical of the type of advice that could be found in contemporaneous popular culture such as sex manuals, women's magazines, or reports on sexuality like *The Hite Report: A Nationwide Study of Female Sexuality* (Shere Hite, 1976) that were circulating as bestsellers. The *Interlude* instruction manual claimed that its scenarios were not only based on research but also drawn from couples interviewed for the program.[30] The imperative toward variety and novelty in the program aligns with what Ehrenreich, Hess, and Jacobs describe as an expansion of ideas about sex to include a wider variety of acts and experiences that had reached the mainstream by the 1980s.[31] In the scenarios it suggested, the program assumed that the couple might have access to such sexual accessories as vibrators, body oils, pornographic materials, and lingerie. *Interlude* offered a selection from 106 sexual scenarios provided, and the manual promised users that "the computer can present you with the perfect Interlude each time you play the game!"[32]

The program's conceptions of what makes a "perfect" interlude largely fell into a few different categories that favored sexual play above intimate communication. Some interludes, like "The Chase," "Hooky Nookie," and "The Wrestling Match," offered ideas to inject play or more adventurous romance into the couple's routine by suggesting they meet at a bar and pretend to pick each other up for the first time, play hooky from work to spend the day together, or attempt an erotic wrestling match. Other interludes were more interested in introducing novel positions and accessories into the couple's sex life, including mirrors, bondage, rocking chairs, vibrators, or sex swings. They also offered suggestions about how to stimulate particular erogenous zones that a user had expressed interest in. Less commonly, interludes were directed to the pursuit of greater emotional sensitivity, awareness, or attentiveness between the couple. "Fantasy Island" encouraged partners to communicate their fantasies with each other, "Feelings!" instructed men to keep a love diary about their partner that they could later share, and a few others offered variations on the advice to spend time together in sensual massages and touching where partners are told to be attentive to each other's physical reaction and responses.

Taken as a whole, the interludes recommended by this program aligned with the post–sexual liberation ideals in which good sex was not equated with one circumscribed activity but could mean different things to different individuals. Within this framework, there is frequent emphasis on variety, novelty, and technique and only to a lesser extent focus on practicing heightened communication and sensitivity to one's partner. *Interlude* was promising a way to use computers to select the best sexual encounters from a large collection. This was part of the promotion for this program as couples software that could serve to distinguish it from a quiz in *Cosmopolitan* or a self-help book. Even though the type of sex advice *Interlude* offered was quite similar to that found in women's magazines, the program purported to use computer technology to optimize its advice.

Yet, *Interlude*'s potential to mediate coupled relationships came not only from its computationally customized ability to introduce novel sexual scenarios but also from the couple's curated experience with the computer. The program was structured to help instigate sexual play. *Interlude* appears to expect some resistance to the playful mode of sexuality represented by the program. To help overcome this, the computer performs playfully as a third actor in the couple's sexual encounter during the interview. After each response users provide, the screen displays playful commentary egging the user on, such as "Now you're talking," "Hot stuff!" or a disappointed "You're no fun." This banter signals that the computer is present as an enthusiastic and active facilitator in the experience, but it also indicates that the program assumed that such encouragement would be necessary to get both partners in the mood to pursue sexual novelty.

Furthermore, *Interlude* appears to assume that members of the couple may have conflicting desires that the program must mediate. If the moods and desires of each member of the couple do not align, the program mediates these conflicts by avoiding transparency rather than encouraging discussion. When two players sit to answer the *Interlude* interview, answers cannot be saved and must be completed in the same computing session, but the male player completes all his questions before the female player is asked to answer. The ordering of questions presumably allows users to answer their questions privately before their partner takes their turn at the computer, maintaining this as a coupled activity even as it segregates the pair. This structure might help avoid highlighting explicit conflicts between partners if they arose, even as it encourages full candor with the computer program.

In fact, *Interlude* purported to help couples not by highlighting discrepancies as the basis for further conversation but by removing discussion between partners from the planning of sexual activities. This is another way that the program worked as a mediator of relationships rather than simply a program to dispense computer-optimized advice. *Interlude* does not report the answers that each user inputs, but instead directs users to specific interludes that may serve as a compromise or solution to disagreements.

For example, if a male player expresses strong sexual desire while his partner reports the opposite, the program may recommend the interlude "Pussy Cat," in which the man is told to give his partner a scalp massage, gently brush her hair and kiss her neck, and proceed further only as she seems to indicate. "Pussy Cat" does not report the woman's preference for cuddling over sex to her partner but recommends an activity directed to him that allows for physical intimacy without necessarily leading to sexual activity. Similarly, if the male player expresses a lack of desire but his partner wants sex, the program might suggest "Peeping Tom," an interlude that encourages her to take the lead in exhibitionist activity and pleasure herself with a vibrator while her partner watches. These interludes are presented as the informed selection made by the computer to reconcile individual desires with the needs of the couple, but the divergences are not reported. Rather than insisting on transparency, the program deemphasized potentially divergent desires of the separate individuals that make up the couple by taking responsibility for the experience. In other words, it negotiated on behalf of the couple.

Considering the male-dominated computer industry in which *Interlude* was created, it is possible to assume that this program, with its limited emphasis on communication, reflects a view from a heterosexual male programming culture aiming to realize masculine sexual fantasies with the help of computers. After all, *Interlude* tries to loosen inhibitions and suggest ideas for novel sexual play so that he will not have to. There is not an abundance of information about how this program was developed, but one disheartening account about its use in a workplace provides some evidence for this reading. In a review of *Interlude* in *80 Microcomputing,* the reviewer offers the following anecdote: "One of our foxiest secretaries here at *80 Micro* displayed so little enthusiasm during one interview session that she was directed to interlude number 29. The scenario for this interlude instructed her to stay home alone and curl up with a good book."[33] In this account, *Interlude* appears to function to pressure this secretary into an unwanted sexually suggestive computer interaction at her workplace.

At the same time, it is possible to read in this same scenario that the program allowed this woman a way to deflect unwanted sexual attention from her coworkers—after all, it was the computer, an objective third party, that decided she should just curl up with a book, thus absolving her of having to express her lack of desire to her coworker. Projecting this reading onto the space of the home—where *Interlude* was presumably targeted and where players of any gender might potentially feel more comfortable cultivating sexual activity—the potential of the program appears ambiguous. On the one hand, it could serve men in realizing their sexual fantasies by helping them encourage women to accede to their desires. More optimistically, though, it could help both members of the couple express desires they might otherwise feel unable to share—whether that desire is for novel sex or for no sex at all.

Even if *Interlude* were not guaranteed to serve both members of the couple equally in its approach to couples computing, it contributed to a broader effort envisioning how computers could mediate coupled relationships and expand their audience. When speaking about the program to general circulation newspapers Syntonic's president, David Brown, emphasized this aspect of the software, suggesting that one justification for a program like *Interlude* was a shift in the market of computer owners: "Computer users are a wide range of people. . . . Look at the average person going into a Radio Shack store to buy computer programs. He's not a 'computer type'—he's just an everyday guy."[34] Although Brown's vision was still confined to a male customer, his comment marked an attempt to expand the computer market and at least imagine how software could be used for couples. Similarly, Syntonic's director of advertising, Sandra Brown, told the computer magazine *InfoWorld* that Syntonic developed *Interlude* as a creative use of computers in response to complaints that microcomputers were only good for "balancing checkbooks." This same article notes the effectiveness of this expanded address, mentioning that *Interlude* was covered in an issue of *Playgirl Couples* magazine, an indication that it had reached beyond the typical computer user demographic.[35] Of course, some promotions and discussions of the game were aimed more directly at men. One computer retailer advertising *Interlude* at their store dared users to purchase this racy software and listed the price for a full-size poster of the "famous *Interlude* girl" from Syntonic's advertising.

By the time a sequel to *Interlude* was released in 1986, the advertising address for this software had shifted. *Interlude II* made explicit appeals to both members of the couple more directly in its advertising. One ad features a scantily clad couple sitting near a Christmas tree with ad copy transcribing their fictional dialogue about the game. The dialogue begins with the woman's initial doubt about receiving a computer program for Christmas but concludes with the couple's mutual enthusiasm about the adventures that await them by playing *Interlude II*. Another ad shows a well (and fully) dressed couple with a bouquet of roses standing in front of a computer (Figure 4). The accompanying text explains that the roses were the computer's idea.[36] Unlike ads for the original *Interlude*, the sequel was promoted more centrally as a way for couples to use the computer together in a manner that would speak to women's fantasies and desires as well.

Interlude was likely the first attempt to create software for home computers to mediate the couple, but other programs followed that placed different emphasis on communication and expression within relationships. Released in 1982 for the Apple II, *Lovers or Strangers* is another example of software that offered couples the opportunity to turn their home computer into a tool for romance. Advertising for *Lovers or Strangers* appeared in computer magazines. However, as described above, rather than making titillating sexual appeals, this program was promoted using imagery that resembled the aesthetics of popular romance novel covers. The ideals it set out for coupled relationships aligned with more expressive modes of intimacy through communication. As suggested by its marketing and

Figure 4. An advertisement for *Interlude II* depicts a couple benefiting from the program. The spot ran in the May 1986 issue of *inCider*.

citations of women's culture texts, *Lovers or Strangers* seemed to have women's desires more centrally in mind than did other romance software. In this program, an ideal couple was one that communicated more, and the software attempted to strengthen the couple's bond by instigating and organizing this communication.

Lovers or Strangers has a complicated production history that was affected by women's changing companionate roles. Alpine Software published the program in 1982 but it appears to have been quickly taken off the market. A rerelease of the game was published in 1984 by Softsmith under the title *Friends or Lovers*. The game is credited to Stanley Crane of Alpine Software, who programmed it. In newspaper accounts from the early 1980s, Crane also credits Brigitta Hillis (née Olsen) for supplying questions and suggestions for the game.[37] In a conversation with the author, Olsen claims to have originated the idea for the game and developed the concept and questions with another partner before asking Crane to program it. Olsen has a unique background within the context of the male-dominated software industry of this period. She holds a bachelor's degree in behavioral psychology and worked as a counselor in sex and sexuality at the commune More House / More University, where she specialized in teaching practices of extended orgasmic states through training in tantric skills. After her time at More House, Olsen worked as managing editor for publications and documentation at a software firm in which Crane was a partner at the time *Lovers or Strangers* was developed.

In contrast to the flippant way many reviews discussed *Interlude,* Olsen claims to have approached *Lovers or Strangers* in part through a sincere desire to use computers to aid women in expressing to their partners their sexual desires safely and comfortably. As she recounts, "I was fascinated with a computer's ability to provide a safe way for men and women to share what was true and authentic about sex for them . . . and what each thought was true for their partners. Talking through the computer took a lot of the charge out of being honest about one's own sexual likes and dislikes."[38] Speaking specifically to social conditions for women at the time the software was developed, Olsen recounts, "We women were now expected to work ('cause that's what we asked for, right?) and have babies and take care of the house and hubby . . . without any recognition or support. It just seemed so important to me at the time that we had a place where we could tell the truth about the most intimate of things—our sexual pleasures."[39]

Olsen's comments and her background suggest that experiences of sexual liberation cultures and feminist critique were finding their way into this computer game; however, this interpretation competes with others regarding the software's origins. According to Olsen, *Lovers or Strangers* was an attempt to use computers to help women and their partners pursue more fulfilling sexual experiences and to negotiate sexuality within shifting gender roles in the home. In newspaper accounts in the early 1980s, in contrast, Crane offers a different explanation of the game's origin. Divorced and wealthy because of his success in the software industry, Crane claimed to have developed *Lovers or Strangers* because he was tired of trying to meet trustworthy women at singles bars who desired him for his personality and not his wealth. According to Crane, the game was a way to make the process of determining compatibility more efficient.[40] Even in Crane's account of the game's origin, Alpine Software hired a woman marketing consultant to help sell the product to consumers, and it was presumably this decision that informed the program's advertising to women.

The emphasis on women's pleasure in one purported origin of *Lovers or Strangers* is reinforced in the promotion for the game that mimicked romance novel aesthetics. Such an allusion was significant not only because it departed from common conventions for software advertising, but because it depended so obviously on a genre associated with women's culture. Romance novels were an explosively popular literary genre throughout the 1970s and early 1980s when this advertisement circulated. Publishers of these novels described their demographic as overwhelmingly women. In her ethnographic work with romance readers, Jan Radway notes that many of the women in the study expressed that romance novels spoke to a desire for escape from the everyday; moreover, she found that readers were attracted to the model of nurturing intimacy represented in these stories.[41] Using imagery from a popular women's genre was another indication that this program was targeting women and their desires. It is possible that Alpine Software was attempting to pique women's interest in computing by aligning *Lovers or Strangers* with fantasies of an escape from everyday modes of domestic communication similar to those found in romance novels. The ad copy for the program suggests that users' relationships could be interpreted as romantic and adventurous: "Were you made for each other? Are the two of you destined for romance? Spend an evening with *Lovers or Strangers,* and find out."

The structure of the gameplay organized by *Lovers or Strangers* closely resembled another feminized popular genre—the television game show, specifically *The Newlywed Game,* which had been a staple throughout the

1960s and 1970s. Like *The Newlywed Game, Lovers or Strangers* creates some separation between users as they answer intimate questions about themselves and their relationship and predict their partner's answers.[42] In her analysis of television programs like *The Newlywed Game,* Mimi White argues that "the couple is dramatized as a split body situated in networks of communication." This split body becomes a couple through their willingness to submit to the programs' imperative toward therapeutic confession. These shows "require confession as part of the process for judging and understanding the couple as a social unit. But simply participating in the show guarantees one's identity as a couple. You do not have to match all your answers or win; you only have to confess."[43] As White argues, these game shows were anchored in an ideal of coupling in which communication and confession were constitutive of the couple.

Lovers or Strangers shared the imperative toward confession and communication from game shows like *The Newlywed Game,* but it used the computer to orchestrate this separation and union in a private computing experience rather than for a TV audience. When playing *Lovers or Strangers,* both members of the couple answer questions simultaneously; for each question they must answer for themselves and anticipate their partner's answer. The program instructs players to share the keyboard between them. The player on the left responds using numbered keys 1–5 while the player on the right uses 6–0. The instructions for the program advise players to place an index card down the middle of the keyboard to discourage them from spying on each other's answers.[44] This workaround reimagines the keyboard, a technology engineered for a single user, as a couples' interface. Such an arrangement ensures that the pair will be brought into close physical proximity when using the computer program. Seated beside each other in front of a small keyboard built for one, users would be more likely to brush up against each other's arms or legs while typing. These moments of physical intimacy could contribute to the excitement and anticipation of the computer interview and draw attention to the two users as a couple. Yet, the division of the keyboard and the requirement to consider how your partner might answer differently also draws attention to the possibility of division within the relationship.

Along with its emphasis on communication, *Lovers or Strangers* treated sexuality as only one topic that couples could discuss in pursuit of greater communicative intimacy. In addition to sexual topics, *Lovers or Strangers* asked about other values such as thoughts on contemporary gender roles. One question inquires if the user thinks it is okay for a woman to make more money than her partner, and another addresses opinions about men and women together in the workplace. As evidenced by these questions, the program was contending with contemporary anxieties about women's accelerated entrance into the paid labor force and the perceived effects this would have on their intimate relationships.

In response to conflicting answers, though, the solution proffered by the program was more discussion. *Lovers or Strangers* scores each user on how well they guessed their partner's answers and tells them in which areas they can improve their knowledge of the other. The program's diagnoses of compatibility were perhaps unnervingly neutral. For example, if users answer the questions in *Lovers or Strangers* in wildly conflicting ways, the program does not recommend that they seriously rethink their pairing. For example, a compatibility score of 41 percent triggers the advice "You two seem to be balanced in your similarities and differences," paired with a dispassionate description of the general areas in which two users might try to learn more about each other. As Olsen recounts, the idea behind this tone was to maintain a sense that the program was supportive and to keep it from getting emotionally charged.[45] Similar to how White describes *The Newlywed Game,* in *Lovers or Strangers* a couple didn't necessarily have to share values or opinions but must be ready to discuss them further.

Lovers or Strangers encourages a confessional form of intimacy between users about a range of topics. Once it has offered compatibility scores, *Lovers or Strangers* encourages players to review each question together and see how their partner answered. These results supply the foundation for ongoing conversation about these issues. As the instructions on screen suggest, "The more you talk about your answers, the more you find out about yourself and your partner." This program was consistent with contemporary discourses of self-expressive intimacy in which the willingness to engage in intensive conversation and confession were constitutive of what it meant to be in an ideal intimate relationship. The communication imperative in *Lovers or Strangers* similarly encourages players to fully disclose answers to each other and to continue conversations about these answers beyond the initial interview.

Interlude and *Lovers or Strangers* were both published early in the spread of computing as ideas about the market were shifting. By the time *IntraCourse*, which was previewed at Comdex in December 1985, was officially released in 1986, the market for home computers had expanded further. *IntraCourse* was presented as an educational program more than a playful or romantic game. In promoting it as a therapeutic tool, the software publisher emphasized the program's debt to contemporary sex research. *IntraCourse* promised to use the computer to maximize the couple's access to information about sexuality.

According to *IntraCourse,* a healthy or optimized sex life is unique to each person. The program encouraged users to submit to the computer's queries and undergo its analysis in exchange for insight about the complexities of their sexuality and their modes of interpersonal relating. In other words, *IntraCourse* promised customization and personalization. As the manual informs users, "The first and crucial step in the investigation of your sexual and interpersonal world involves the collection of information from several areas of your life. Here the *IntraCourse* program acts as both a probing interviewer and an accurate data collector. By asking questions which range from the broad to the specific, from the casual to the intimate, and the personal to the interpersonal, the *IntraCourse* profile assembler is able to gather the essential ingredients necessary to unlock your sexuality and unique style of interacting with others."[46] According to *Intra-Course,* to be a strong couple, partners first needed sexual self-knowledge. In focusing so much on sexual customization, though, the program risked highlighting the desires of the individual over their role in the couple.

IntraCourse was not primarily focused on sexual play or intimate expression between partners. Rather, it emphasized its relationship to more established lineages of sex research and sexual self-help. The packaging of the program reinforces this association through its resemblance to the imagery used in sex manuals like *The Joy of Sex* (Figure 5). Intracorp even used a faux pebbled leather material for the product packaging, adding to the impression of upscale respectability and legitimacy.

IntraCourse consisted of two modules that could be bought separately or together. The first offered sexual analysis for the individual user and the second produced reports on compatibility. *IntraCourse* retailed for $99.00 for the combined analysis and compatibility modules, making it significantly more expensive than *Interlude* and *Lovers or Strangers,* which sold for $19.95 and $29.95, respectively. *IntraCourse*'s analysis was based on answers to an extensive questionnaire consisting of fifty to one hundred questions. The program highlighted the "comprehensive" interview that covered a range of topics. Questions asked about number of previous sexual partners; thoughts on masturbation and threesomes; and frequency of oral sex, anal sex, and participation in sadomasochistic practices. Some questions addressed relationships more abstractly, asking if the user feels

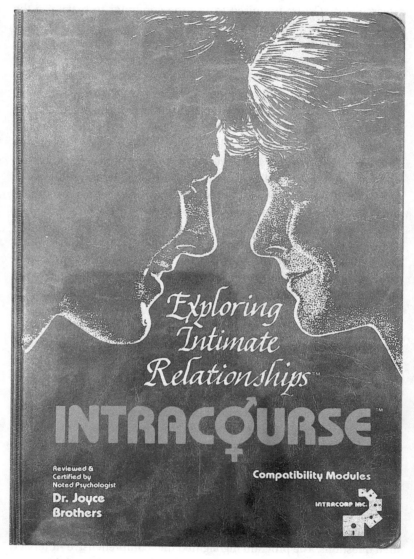

Figure 5. Front cover of *IntraCourse* software packaging, produced by Intracorp, is made to look like popular sex manuals.

sharing with a partner may lead to a loss of individuality, if they believe themselves to be capable of "real intimacy" with a person they just met, if they prefer to withhold information that might hurt their partner, and how they characterize their childhood.

Marketing for *IntraCourse* emphasized the program's proximity to professional psychological knowledge and traditions of sex research. According to Intracorp, the company created the program in consultation with over a dozen psychologists. As one advertisement claims, "IntraCourse improve[s] your relationships and compatibility with information and research from more than 100 sources and studies in human sexuality—including Kinsey and Masters & Johnson."[47] The program also came with an endorsement from Dr. Joyce Brothers prominently displayed on the software packaging. Brothers promised that *Intracourse* "provides the public direct communication of information which previously was available only to behavioral researchers and clinical practitioners."[48]

As a professional psychologist, Brothers's involvement in the program linked *IntraCourse* to professional discourses of sex research but also to women's culture. She had been a staple on radio and television since the 1950s and a prominent figure for women audiences. In addition to noting her broadcasting career, the Brothers bio provided in the *Intracourse* instruction manual also referred to her column in *Good Housekeeping*, her frequent appearances in national newspapers, and the fact that popular polls had voted her "one of the ten most influential American women"; it also noted that she was a wife and mother.[49] As Leigh Goldstein argues in her discussion of Brothers's tenure on television, Brothers has long served as an expert who addressed and counseled women about relational problems, including sex's place as a significant part of companionate relations. Goldstein argues that on her television program, *Dr. Joyce Brothers*, she created a style of discussing sexual problems through an "intellectualist" mode in which "better sexual relations can be achieved, but it requires a lot of thinking and some carefully premeditated conversation."[50] The style of discourse about sexual relating in *Intracourse* closely resembles this mode. Brothers's endorsement reflects a clear attempt to attract women to the program. Her involvement also indicates the impact of her mode of sex talk on the program and the influence of researchers like Alfred Kinsey and William H. Masters and Virginia E. Johnson, who were also cited in the program's bibliography and, as Goldstein notes, influenced Brothers as well.

The associations with professional therapeutic culture were likely part of Intracorp's attempts to justify the program's high purchase price. However, these citations to figures like Brothers and Kinsey also influenced how the program addressed users and discussed sexuality. For example, one of *IntraCourse*'s many features was a statistical analysis. After completing the computer interview, users could run an analysis of their responses to see how their own stated sexual behaviors and preferences compared to national norms, as determined by the sources consulted in the program. A woman might report her frequency of masturbation and see how this aligns with reported national norms. The *IntraCourse* manual suggests that the purpose of this analysis is to destigmatize these results by answering the question "Do other people have preferences similar to my own?" The manual also instructs users how to make sense of these results telling them not to look for negative or positive judgmental statements, but rather to seek unique information about themselves and their relationships.[51] The analysis also offered references to further reading that pulled from popular sex research like that from Hite and the studies of Masters and Johnson.

Relying on contemporary sex research, the program accommodated users who wanted to explore information on a wider range of sexual practices. *IntraCourse* departed from other examples of romance software by including the option to identify as homosexual, bisexual, or asexual and to report varied sexual orientations. *IntraCourse* also included a dictionary of sexual terms that users could access on the computer—including definitions of anal intercourse, coitus, gender role, homophobia, butch, and closet queen. This feature allowed the program to claim a more encyclopedic coverage of sexuality and appear more serious and legitimate as a tool for education. *IntraCourse* imagined computing as an activity that could strengthen the couple, but it first encouraged sexual exploration for each user individually.

IntraCourse incorporated changing ideals about sexuality and intimacy to present a more expansive program for couples. But this also posed some challenges to its role as a technology to mediate coupled relationships. For example, one feature that Intracorp promoted as a sign of its value was its use of an adaptive questionnaire. The software manual highlighted this feature as evidence of the program's customization and personalization, promising hyperbolically to cover "the spectrum of sexual phenomena."[52] Answers to earlier questions in the interview triggered relevant follow-up questions on sexual habits. One interview might include multiple questions about the frequency, nature, and type of fetish objects that the user includes in sexual activity. This line of questioning results when a user previously answers affirmatively about fetish use. In another interview iteration, the program may not include this line of follow-up questioning. This means that even if a couple uses *IntraCourse* for a compatibility report, they are not encountering identical questionnaires in their initial encounter with the program.

The program's attempt to customize its advice and unlock the truth of sexuality had the potential to work against the coupling imperative, instead offering each user a more personalized sexual journey. Completing the entire *IntraCourse* interview for a single user takes a significant amount of time and involves answering explicit questions. Because an interview session could take so long and could be saved to disk, this part of the program was likely intended as a more individualized and private encounter between one user and the computer, rather than an interview to complete simultaneously with one's partner. This sense of an individual, private encounter with the computer is reinforced by the questionnaire itself adapting to the responses each user provides.

Of course, as compatibility software, *IntraCourse* was not entirely personalized, and the program encouraged sexual exploration to strengthen the couple. *IntraCourse* required users to save their answers to a personal profile that would be labeled with a chosen username. Users could password protect their profiles if they desired.[53] *IntraCourse*'s password protection allowed users to restrict access to their profiles, but passwords were also meant for sharing with potential matches. In fact, to access the compatibility feature in *IntraCourse,* users were encouraged to obtain the password of the partner with whom they wanted to be compared. In this case, password protection was not intended purely as a way for an individual to keep their exchange with the computer private. Instead, this was a selective privacy in which passwords were exchanged as an act of intimacy between users. The compatibility profile offered quantitative compatibility scores in twelve categories—such as dominance/submissiveness, independence/dependence, and strong/weak sex drive. *IntraCourse* also offered a narrative compatibility report to help the couple make adjustments to compensate for differences in their personalities, sexuality, and relationship styles.[54]

Interlude, Lovers or Strangers, and *IntraCourse* all offered different models of how the computer might be used as a technology to mediate coupled relationships. It might playfully instigate users and help negotiate their conflicting desires, as in *Interlude.* It could also work to prompt expression about one's desires, including sexual desires, as a precursor to better communication with a partner, as in *Lovers or Strangers.* Or alternatively, as in *IntraCourse*, the computer could function as an educational tool to guide the user through sexual exploration to facilitate more optimal coupled interactions. As well, the programs aimed for different ideals of coupling, emphasizing novelty, expression, and exploration to varying degrees. Yet, although they worked differently, *Interlude*, *Lovers or Strangers*, and *IntraCourse* were all influenced by discourses about couples circulating in popular culture, sex research, and self-help literature in the 1980s.

"An On-Going, Invaluable Gauge of Your Sexual Self"

Romance software promised to maximize the pleasure of computing for couples. However, advertising, promotional materials, and reviews of these programs reveal some of the contestations and tensions involved in using the personal computer to help couples sustain their relationships. These debates questioned whether these programs prioritized men's or women's desires in a heterosexual couple, and how they would relate to monogamy or alternative arrangements. Would romance software be used solely by married couples or might it encourage other forms of intimacy or sexuality? These programs were purportedly designed to balance individual desires and fantasies with the needs of connection and intimacy in committed relationships, but would they encourage users to explore individual fantasies at the expense of the monogamous couple? These questions revolved in part around whether these programs were serious software products or just games to entertain users.

At times, promotional material and advertising for romance software emphasized that the programs were aimed specifically at existing couples as a playful but functional tool to inject novelty into sexual relationships. The instruction manual for *Interlude* promises, "The *Interlude* program can help you discover or rediscover the playful, the exciting, the erotic qualities in your relationship, as well as revive the romance and tenderness that may have slipped away unnoticed."[55] The reference to reviving and rediscovering romance implies that the program is for couples who have been together for a while, not for new or future couples. Still, another advertisement for *Interlude* hints at an alternative to the couple when it describes the program as "a computer game that turns your love life into a *ménage à trois* ... you, your mate, and your computer!"[56]

Discussions of *Lovers or Strangers* also showed an ambiguous relationship to the monogamous couple. Upon the program's rerelease under the title *Friends or Lovers*, the vice president of marketing for the software publisher highlighted the involvement of two psychologists in its development: "It is not just some programmer sitting down to write it. . . . Two psychologists wrote it and flushed out the core of interpersonal issues." According to those involved in its production, "the program wasn't just a game, it was a different genre—psychological software."[57] This appeal to professional psychological authority attempts to shore up the program's value as an accurate way to test compatibility. As discussed above, Crane claims the program was a way to test compatibility with women more efficiently than a singles bar. He even claimed that he married a consultant on the program after playing *Lovers or Strangers* with her and that their marriage was still going strong.[58] Although this story purports to assert the legitimacy of *Lovers or Strangers* as a tool to promote marriage and monogamy, it also hints at the ambivalent role the program could play. Not only could existing couples use the program to strengthen their relationships, but Crane's testimonial suggests that a consumer could use *Lovers or Strangers* to quickly test out compatibility with a slew of different partners. Still, even in this more fantastically promiscuous scenario, Crane's story suggests that the end goal of the program is a long-term and fulfilling marriage.

Intracorp used a variety of methods to promote *IntraCourse*, both emphasizing its relationship to professional research and hinting at more titillating experiences. One advertisement promises the program will "unravel your wildest fantasies." This spot describes *IntraCourse* as "the 'sexy' software" and promises users "high-tech sex."[59] At the same time, Intracorp's advertising also indicated that the program was aimed at couples. Intracorp distributed the software through mail order advertised in computer magazines and retail stores like Sharper Image, evidence that the program was targeted at a wider market. *IntraCourse* was even promoted in a Sharper Image insert in *Popular Science* along with an advertisement for a talking bathroom scale.[60] Although this magazine tends to target male hobbyists, the advertising imagery suggests that *IntraCourse* and the adjacent scale are also meant for women—the bathroom scale is staged next to pink ballet flats and the software is accompanied by a single red rose. The advertisement targets men who might buy the program to use with a partner, with the red rose hinting that it could replace or supplement a romantic gift of flowers.

Romance software appeals to the legitimacy of seemingly scientific psychological approaches, connecting them to the history of psychological testing and the field of marriage counseling. Rebecca Davis explains that anxieties about difference underwrote the earliest scientific efforts to mediate the couple dating back to at least the 1930s. Often informed by eugenic agendas, marriage counseling encouraged unions between educated, upper middle-class white Americans. Davis argues that in subsequent decades, marriage counselors turned to information technologies and quantification to reinforce traditional gender roles and the belief that long-term relationships are based on social similarity.[61]

Like marriage counselors' embrace of information technology, romance software supported the couple by using computer technology to improve their romantic and sexual encounters. Notably, though, romance software pursues this goal through a more flexible approach to sexuality in which the programs process information about sexual desires and preferences to enhance pleasure. They articulate an ideal that more closely resembles what Giddens, as described above, has referred to as the "pure relationship."[62]

Some critical reviews expressed these tensions by speculating on how these programs might be used outside the monogamous couple. A *PC Magazine* feature listing computing trends opposed by the magazine's editors referenced *Interlude II* as an example of a misstep in the development of computing.[63] Although the editors do not explain why *Interlude II* was selected, they jokingly assure readers that the woman featured in the *Interlude II* advertisement in a previous issue is wearing a wedding ring, thus presumably legitimating her sexual activity hinted at in the spot. The disclaimer implies that some people may not like the program because they think it encouraged promiscuity or sexual relations outside marriage.

The popular computer magazine *Creative Computing* featured a review of *Lovers or Strangers* in its 1983 "Buyer's Guide." Rather than labeling it "psychological self-analysis" like another program reviewed in this issue, the reviewers classified the software type as "Game?," the question mark highlighting their uncertainty about how this software might be used. The reviewers acknowledged that they played it on a break from writing and tested the compatibility between two long-term coworkers rather than a traditionally romantic pair.[64] Another review of *Lovers or Strangers* proposes that it could be used as a party game to entertain guests.[65] This suggestion focuses more on the entertainment value of the program than its utility to existing couples.

Additionally, advertising for *IntraCourse* framed it at times as a repeatable play experience. Promotional material for the program claims, "As you become more adventurous and creative, with new desires (and/or partners), you will find *IntraCourse* an on-going, invaluable gauge of your sexual self."[66] Here Intracorp suggests that the program and the sexual experiences it offers might outlast the couple in question and instead be used for a range of partners. Reviews of the program also toyed with this idea. In one computer magazine, a reviewer suggests that the program's greatest value is in entertainment and more speculative ideas about coupling: "If you'd like to sit down with a group of friends and let the computer suggest who 'really' ought to be paired up with whom, you could have some fun. Or spend an evening by yourself trying on personalities and seeing which ones are the most interesting."[67] In fact, *IntraCourse* had the capacity to store twenty different user profiles. This feature enabled each user to create multiple profiles or test multiple novel pairings, which could spur different relationships between individuals who already knew each other. Users could approach romance software as repeatable programs and even as computer games. As games, they had replay value and encouraged the user to engage, at least speculatively, with multiple partners or multiple ideas of their own sexual desires and preferences.

As computers entered domestic space in the 1980s, some software companies were envisioning how the devices could be used to mediate, rather than simply distract from, relationships and marriages. Romance software represented a unique way of thinking about computing technology in the home. Computers could function as mediators as couples negotiated normative conceptions of intimacy, romance, and sexuality. Even when romance software addressed male users or was developed by men, it was shaped by women's culture and feminist contestations over gender roles and sexuality in the companionate family. What it meant for the computer to function as a tool to mediate the couple, then, was being defined in significant ways by these cultural forces.

Historians of technology have discussed some of the ways computing intersected with relationships and sexuality in prior periods before the spread of computers into the home. For example, in Mar Hicks's analysis of computer dating in the 1960s, they argue that although these services were often promoted as novel and revolutionary, they were in fact "predicated upon reinscribing conservative social norms into a new set of technological systems" by matching like with like and assuming heterosexual marriage as the ultimate goal.[68] Alternatively, Donna J. Drucker has illuminated the history that information technology, in the form of punch card machines, played in Alfred Kinsey's reports on sexual behavior, which Drucker argues was "instrumental in developing gay and lesbian rights movements."[69] These studies and Drucker's argument suggest that the computer's role in heterosexual relationships cannot be reduced to its ability to rationalize information about relationships and sexuality or process large sets of data. As these scholars show, computers have had an ambivalent relationship with histories of sexuality and coupling. Like these other accounts showing how computers have either maintained or reshaped heteronormative coupling, my analysis of romance software in the 1980s shows how computers were implicated in contemporary contestations regarding intimate and sexual bonds. With romance software, though, computers functioned as a medium of romantic relationality by orchestrating and shaping how couples interacted rather than by influencing conceptions about who should be joined in intimate connections.

2 "NOT AN APPLIANCE, BUT A FRIEND"

Personal Robots and Participant Fatherhood

IN 1984, THE HEATH COMPANY BEGAN SELLING a personal robot called Hero Jr. An advertisement for the robot implored readers, "Take your family beyond computers. . . . Enter the fascinating world of the home robot." The advertisement goes on to describe Hero Jr. as a friendly companion and describes some of its features: "He'll wake you in the morning, guard your home at night, and entertain throughout the day with small talk, songs, and games." This spot and similar ones promoting Hero Jr. circulated in computing and electronics magazines in 1984. Heath's promotional address is aimed at the typically male readers of these magazines, but it implies that by buying Hero Jr. this reader can bring his family along as he explores novel technology. The ad concludes, "There's a lot to learn . . . and even more to enjoy! So, for less than the cost of a home computer, introduce your family to the wonders of robotic living."[1] With Hero Jr., this advertisement promises, the user can learn about robotics and programming while benefiting his family at the same time.

In the early 1980s, a few companies developed and sold personal robots for the home. These included firms with established records in electronics, such as Heath, a subsidiary of Zenith, as well as new outfits founded specifically to pursue robotics, like Androbot and Hubotics. Although not a typical desktop microcomputer with keyboard and screen, these personal robots still represented a type of computing and were associated with the developing computer culture of the 1980s. Many of the designers and engineers involved in producing personal robots had experience in the microcomputing and video game industries, and the robots were powered by the same microprocessor technology used in desktop computers. In fact, commentators described them with phrases like "a personal computer that walks around" or an "IBM-lookalike-on-wheels," framing them as the next stage of the computerization of everyday life.[2] In addition to being computers in their own right, most robots, including Hero Jr., could be connected to a desktop microcomputer to facilitate their programming and use. Thus, personal robots were also described as peripherals or accessories for already-purchased home computers.[3]

And yet, although personal robots were an application of home computing, these technologies were also explicitly framed as robots. When discussing these new devices, rarely did a newspaper or magazine article refrain from referring to R2-D2 and C-3PO from the *Star Wars* film franchise.[4] Just as frequently, these articles contextualized the new technology as a step closer to achieving long-circulating fantasies of robot butlers and electronic domestic servants—although they warned that the realization of practical tasks was still at least a decade away.[5]

Personal robots carried expectations associated with these dreams of labor-saving domestic appliances and technologies. Like other robots and electronic appliances in the past, personal robots in the 1980s were presented as technologies to support and enrich middle- and upper middle-class family life. Drawing on the long history of promoting domestic devices as assistants to women, personal robot manufacturers occasionally gestured to the way robots could act as women's helpers or companions in the home. But unlike many fantasies of electronic domestic servants before and after, these personal robots were largely targeted to male hobbyists. As such, in working to fit these technologies into the home, manufacturers and commentators situated these personal robots through appeals to images of masculine domesticity. As the advertisement for Hero Jr. suggests, these home robots were presented as technologies that men could use with their families or in service of family life without having to sacrifice their computer hobby.

Analyzing the promotion and design of personal robots from Heath, Androbot, and Hubotics, as well as discourses circulating about these devices at the time, this chapter argues that personal robots in the 1980s were framed as tools to assimilate a male hobbyist fascination with computers and programming to ascendant cultural expectations for men's increased participation in fatherhood and domestic affairs. Participant fatherhood, as it was developing as an ideal in this period, required that men spend more time with family and children and integrate themselves into domestic life. Personal robots were presented as tools to extend and support men's ability to relate to their families, especially their children, and fulfill expectations for paternal care. According to advertisements, fathers could program robots to entertain children or provide educational experiences for the whole family. By learning to use these robots, it was promised, fathers could gain familiarity with robotics and programming in ways that would integrate them into their family relationships rather than alienate them from domesticity.

This emerging ideal of participant fatherhood also came tasked with expectations for a more affectionate and nurturing mode of relationality with children. In addition to marketing these robots as ways for fathers to extend their parental and familial attention and care, personal robots were designed to assimilate into domestic spaces as cute pets or children. In this way, their design also encouraged a more nurturing form of computer interaction.

As this chapter will discuss, home robots promised to extend and support participant fatherhood in these two primary ways—they extended men's computing interests into family-oriented activities or as acts of care for the family; and they also supported users as they learned to care for cute, vulnerable robots, which served as training in the type of affectionate and nurturing paternal care idealized at the time. Yet, robot makers' attempts to present these devices as advantageous to family life was complicated by technological limitations and the realities of domestic space. Robots had little practical use, and their cute design and painstaking programming threatened to absorb men in robot care rather than childcare.

Home robots were marketed primarily to men, but their design and marketing show the influence of women's domestic culture and of feminist critiques shaping new ideals of participant fatherhood. As devices for masculine domesticity, they still engaged with histories of representation of domestic appliances often associated with women. Furthermore, when robot marketing addressed men, they were not hailed as isolated hobbyists or tinkerers. Robot promotion took into account how men fit into their familial context, a role that was shifting as women's domestic and public roles also changed. These discourses considered how men's robot use also affected their families.

When computers began to spread into U.S. homes in the late 1970s and 1980s, the programming labor necessary to operate these machines was often depicted as a masculine type of pursuit, separate from everyday companionate family life; learning to program could help men pursue a technical hobby or extend the productivity of public workplaces into the private spaces of their home. Although not commercially successful, personal robots provide an important counterpoint to this vision of programming labor and robot tinkering. They imagined this work as a domestic activity associated with affection and caretaking and that could help men balance their interest in programming with their role as nurturing fathers.

Participant Fatherhood in the Computer Age

In the late 1970s and 1980s, it was common for companies to promise that computers could help extend the economic productivity of public workspaces into the domestic sphere. Elizabeth Patton argues that computer use at home was associated with upward mobility through work but presented in a manner that integrated work with family life. Especially for professional men, "home-based labor meant individual freedom and work–life balance without sacrificing income potential."[6] As Patton notes, computers and telecommuting were part of the imaginary that made this possible—enabling productivity from home while allowing men to be closer to their families.[7] Hints of this promise can be found even in the earliest print ad introducing the Apple II in 1977. This image shows a man using his Apple II to work from home, signaled by the line graphs displayed on the monitor in front of him. His work is carried out on a kitchen table in proximity to his wife, who prepares a meal while looking on at his work. Engaged in different types of labor, the couple is nonetheless united in the home. Although the image does not include children, the adjoining copy suggests how a user can involve them in this new computer purchase: "Use it to teach your children arithmetic, or spelling for instance."[8]

As media historians have pointed out, there are many reasons why this picture of productive computing from home and integration of labor and family may not have been as simple as it looked in computer advertising. For many women, even if personal computers were available to them, it was difficult to incorporate working from home with the responsibilities of domestic life.[9] Even for professional men who may have found the balance of family and work more manageable, the computer was still a novel, complex technology. It often required labor and deliberation to make it productive, which could distract men from family time.

Still, despite the challenges of using computers to integrate work and family life, discussions in both specialty and general interest periodicals featured stories that latched on to these promises. For example, in July 1983, *Parents* magazine ran an article in which a father described his family's first year with a new Apple II computer. In the piece, "A Computer in the Family," James A. Levine describes his family's changing relationship to computing as they used it for work, education, and entertainment.[10] When he first brought the computer home, only his seven-year-old son was excited, whereas his wife and daughter showed no enthusiasm. Levine goes on to describe how the computer was integrated into their home, emphasizing how his purchase of the Apple II served his family and made him a better parent. Upon buying the computer, Levine fantasized about his kids turning into tech whizzes. He purchased computer games for his children and watched as they became more competent players, but he worried they would be influenced negatively by game violence. He regularly played computer chess with his son after dinner and observed as his wife and children took to word processing.

Throughout the article, it becomes clear that Levine is not a typical computer hobbyist. Rather, he is emblematic of the wider audience computer companies were trying to reach in the 1980s. Levine bought an Apple II after reading articles proclaiming that children must learn to program in order to succeed in future job markets. He assumed his family would use it for this purpose, but he was not familiar with any programming languages. Levine explains that he began to teach himself BASIC and then Logo out of a sense of responsibility to his children: "Sometimes I feel guilty that I haven't learned more about programming, as if somehow I'm not adequately preparing my children for the high-tech future."[11] Only near the end of this six-page account of his family's growing relationship to computing does Levine address his own individual attachment and pleasure in computing. He explains—perhaps defensively—that his family has chided him for talking about the computer too much, even though he does not believe he has an unhealthy fixation. Finally, after summarizing again how far his wife and children have come in relation to computers, he ends by declaring himself a proud parent—although he is referring not to his son and daughter, but to his computer "baby."

Levine's account of bringing a computer into the family demonstrates some of the ways popular discourses were imagining how general users could employ computers as part of their labor and leisure in the home and how this was deeply tied to their family relationships. He uses the computer to write his article, which seems to supplement his primary job, and delights in his growing facility with the technology. But the computer is also deeply integrated into his relationships with his wife and children. Through using the computer together and teaching his children about programming, Levine attempts to perform his parental care—spending time with his kids, making decisions about their upbringing, and helping secure their futures. Levine thinks he will be a better father if he can learn to program and pass this ability on to his children. He uses the computer to bond with his family, but occasionally his fixation works at odds with this, as he seemingly irritates his loved ones with his nonstop talk on the subject.

Levine's article shows the challenges of integrating computers into family life, but it also offers a glimpse into the changing ideals of fatherhood in the 1980s, when men were expected to be more involved in parenting. It also hints at how computers could be useful in negotiating these shifts. Levine shows a keen interest in child-rearing; he is not just a breadwinner but also actively involved in home duties and play. The computer helps him in both these roles. In fact, when he wrote this article for *Parents* magazine, Levine was director of the Fatherhood Project at the Bank Street College of Education in New York and had written a study of fatherhood and child-rearing. He was a figure in the feminist men's movement that began in the 1970s as a response to and an extension of women's liberation. Kirsten Swinth has argued that although this men's movement was never large in number, it had wider cultural influence as ideals for fatherhood were being renegotiated in the 1970s and 1980s.[12] The work of this men's movement toward revising norms of masculinity and fatherhood continued into the 1980s and included fathering classes offered at Bank Street College. Levine's experience as a fatherhood specialist and later a consultant for companies like Apple helps demonstrate how changing ideas about men's labor and their place in the companionate family influenced the way computers were taken up in the home.

Like Levine, many men in the 1980s saw their work lives and family roles changing, with computers often playing a part. At this time, the United States was increasingly shifting toward an information economy, continuing a process of deindustrialization that began in previous decades. Companies disinvested from manufacturing, outsourced jobs to lower-paid labor forces outside the United States, and continued efforts to depress the value of U.S. labor and weaken unions.[13] Industrial automation and computerization were identified as major economic and technological forces that contributed to this process. The era of industrial robotics was inaugurated decades earlier when the Unimate was installed at a General Motors plant in 1961, but by 1970 there were only two hundred industrial robots in use in the United States. By 1980, this number had jumped to over four thousand, and it was estimated that fifteen hundred robots would be adopted in American factories in 1981 alone.[14] As a result, the early 1980s was a period of intense discourse about robots, computers, and their economic impacts. Articles on the "robot revolution" could be found widely in publications like *Time* and *Newsweek* and were the topic of multiple congressional hearings and reports.[15] The topic of this chapter is personal robots for the home, but the marketing of these consumer devices occurred within this larger context of industrial automation, computerization, and changing labor circumstances.

The adoption of computers and robots into workplaces and industry was identified with the changing value placed on labor in the ascending information economy. Many of these economic shifts negatively affected the working class, leading to higher unemployment and depressed wages. In the 1980–81 recession, for example, there were three blue-collar jobs lost for each one white-collar job.[16] Yet, middle-class families, those most likely to adopt computers and robots for the home, were facing greater insecurity as well. Membership in the middle class was shrinking: the proportion of middle-income families decreased significantly, from 53 percent in 1973 to under 48 percent in 1984. At this time, a household with only one middle-income earner was no longer guaranteed a middle-class lifestyle. This put increasing pressure on members of the professional middle class to seek out stable careers. Barbara Ehrenreich argues that this was part of a "yuppie strategy," in which young members of the middle class pursued pragmatic college degrees and careers in fields like business, finance, and engineering to help secure their class status.[17] Although she doesn't discuss computing directly, training in programming and robotics were framed as lucrative options that could expand economic opportunities or inoculate individuals and families against the threat of downward mobility. Computer and robotics companies relied on this promise to make their products more appealing to potential consumers.

Changes in economic and labor patterns reverberated into family life. The traditional family model with the father as the single breadwinner was becoming less common. Women, including those with children, entered the workforce at unprecedented levels, and many adults found themselves part of two-job families. More women entering the paid labor force meant more families renegotiating domestic roles. In her study of such dual-income families, Arlie Hochschild found that the reality was that men often did not pick up an equal amount of housework to compensate for their wives' new positions, which put pressure on family relations.[18] Despite that finding, the expectation that some men would take an increasing role in the home gained ground in popular discourse.

Norms for masculinity and fatherhood were put under pressure by both the economy and changing social conditions encouraged by feminist movements. For example, the feminist men's movement, which had begun in the 1970s, continued into the 1980s. In 1982, the National Organization for Changing Men was formed. Workshops and consciousness-raising groups helped men deconstruct masculine norms as well as increase their emotional awareness and understanding of feminism and gender roles.[19]

Fatherhood was a significant facet of changing norms of masculinity during this period. E. Anthony Rotundo argues that in the late 1970s and 1980s a new ideal of participant fatherhood emerged, characterized by fathers being actively involved in their children's daily lives and childcare duties. Participant fathers see their duty to their children as wide ranging and less segregated by gender. They are expected to be expressive and intimate with their children.[20] Organizations like the aforementioned Fatherhood Project attempted to train men in more nurturing forms of parenting, teaching them how to be more involved in their children's lives.[21]

Participant fatherhood, a conflicted ideal, was not the reality for most families. As Rotundo notes, only a small fraction of fathers fit with this style of parenting. Participant fatherhood was in many ways a privilege of upper middle-class households. Job responsibilities and inflexible work hours meant that many men, even if they desired to be more involved with family life, did not have the means to do so.[22] For other men, even when the possibility of participant fatherhood was economically available, they felt these increased expectations posed a risk to career advancement or might impinge on their leisure interests and personal freedom.[23] Additionally, fathers who were able to be more present in the lives of their children sometimes lacked the emotional skills or experience to sustain the intimate involvement expected of them.[24]

Nonetheless, in this period, some families tried to address shifting gendered expectations by experimenting with different models for labor inside and outside the home—often with the help of computers. *Family Computing* magazine reported on a single father of two and laid out the different software he used to streamline his care of the home so that he could balance his domestic time with his work as a teacher. He used spreadsheet software to equitably assign chores and plan meals and grocery trips, boasting that the computer saved him so many hours that he could better spend bonding with his children.[25] In an article on telecommuting,

InfoWorld profiled a father who worked from home while his spouse continued to go into work every day. This father reported on the challenges of separating work and family that emerged from this arrangement; for instance, he had to resist the temptation to have his children answer phones or take messages. The author described the resulting setup as "a role reversal," in which the father prepared meals, drove his children to school, and cared for them when they were sick, all while his wife was away at work.[26] Yet, even though computers enabled some men to work from home, they did not necessarily lead to more equal gendered expectations for childcare or greater participation of men as participant fathers. One article reported on a work-from-home employee of the Control Data Corporation who sent his daughter to a neighbor during the workday because, as he reports, "I was there to work, not to baby-sit."[27]

The conflicted ideal of participant fatherhood was part of redefining men's role in domestic life and the companionate family. But this was not the first time such changes had occurred. Contestations over participant fatherhood in the 1980s are part of a longer history of masculine domesticity. Margaret Marsh locates the emergence of masculine domesticity in the late nineteenth and early twentieth centuries. In relation to that period, she defines masculine domesticity as "a model of behavior in which fathers would agree to take on increased responsibility for some of the day-to-day tasks of bringing up children and spending their time away from work in playing with their sons and daughters, teaching them, taking them on trips."[28] As she describes, this model was still far short of a feminist masculinity. It did not presume that men would share household duties equally or that women would have greater access to opportunities outside the home. Still, men were more involved than they had been in previous generations. In the 1980s, expectations for men's participation in work and child-rearing were changing again, but these shifts did not necessarily lead to more equitable gender roles.

Attempts to make sense of changing fatherhood ideals and men's domestic roles were addressed in a variety of popular media. TV scholars like Alice Leppert and Bridget Kies have discussed how a more sensitive masculinity was circulating on television sitcoms of the 1980s. Television offered images of men doing domestic labor alongside women and even excelling at what was historically defined as women's work.[29] Kies, for example, discusses television's "Mr. Mom" characters that stood in contrast to "1980s Reagan-era machismo." In sitcoms such as *Who's the Boss?* (ABC, 1984–92), *Charles in Charge* (CBS, 1984–85), and *Full House* (ABC,

1987–95), Mr. Mom characters demonstrated that men could be effective caretakers and, it was suggested, might even replace women altogether. Yet, like masculine domesticity in the early twentieth century, such developments in masculine norms were not necessarily feminist. Kies argues that depictions in these sitcoms of sensitive, nurturing men often came at the cost of regressive images of wives and mothers, who were shown to be unnecessary to the functioning of the home.[30]

Representations of men on television provided fantasy images of households in which labor was shared equally and harmoniously. But, as Leppert argues, these sitcoms were part of a network strategy to target working women.[31] In contrast, personal robots were technologies aimed primarily at male consumers. These robots for the home supplied their own fantasy visions of masculine domesticity, catering to the desires of a presumed market of men. Robot companies marketed personal robots to men as tools to help them manage their greater involvement in domestic life. They often did this by speaking to men's interest in computer programming and robotics, offering visions of how this could bring men closer to their children and help satisfy their family responsibilities.

Personal Robots and Men's Changing Domestic Roles

Robots were pervasive figures in the 1980s, associated both with fantasies of high tech as well as anxieties about changing labor due to automation and computerization. Science fiction and fantasy images of robots and their risks to society could be found in films like *Blade Runner* (dir. Ridley Scott, 1982) and *The Terminator* (dir. James Cameron, 1984). Features on both industrial and personal robots appeared in computer magazines as well as general interest publications. Magazines like *Robotics Age* were published to keep hobbyists updated on new technological developments. In 1984, enthusiasts organized the First International Personal Robot Congress, which included a remote address by science fiction writer Isaac Asimov. This gathering brought together consumer robotics companies, amateur roboticists, and even representatives from industrial robot firms. Robots were demonstrated performing a range of applications such as sensing and extinguishing small fires or mowing lawns.[32]

A few companies tried to exploit this interest in robotics by producing commercial robots for personal use. Heath's Hero 1 and Hero Jr. robots, Androbot's Topo and BoB, and the Hubot from Hubotics are representative of the various strategies used by robot manufacturers to market to the home. Of course, these personal robots were not the only ones being produced in this decade. Other home robots were sold, such as the RB5X by RB Robotics and the Genus from World of Robots. Toy companies were also selling robots targeted primarily to a children's market. These robots, like Ideal's Maxx Steele and Tomy's Omnibots, were programmable and some even used microprocessor technology. Although such products often had impressive capacities, they were developed primarily as toys, so they were not promoted as a useful part of everyday life in the same way as the personal robots discussed in this chapter. Amateur roboticists were also active in this decade. Books with titles like *How to Build Your Own Working Robot Pet* (Frank DaCosta, 1979) and *The Everyone Can Build a Robot Book* (Kendra Bonnett and Gene Oldfield, 1984) described innovations in robotics and detailed what was needed to build one's own unit at home. Robots made by amateurs and tinkerers for their own use also showed how computing could fit into everyday life, and many of these visions were incorporated into commercial ventures. Yet, unlike these amateur efforts, commercial robot firms consistently attempted to make home robotics popular. These companies labored to persuade consumers that personal robots were necessary and desirable for harmonious domestic relationships.

But what role such robots would play in the home was not immediately obvious. The firms that designed and marketed personal robots throughout the decade did not rely on a consistent strategy for promotion. In fact, the first personal robot discussed in this chapter, the Hero 1 by Heath, was primarily marketed to trade schools as a training tool in robotics; the home market was a secondary concern. Nonetheless, a strategy emerged of framing robots as a support and extension to men's labor and attention in the home, especially with regard to child-rearing and participant fatherhood. Heath followed up the Hero 1 with a robot marketed as a domestic pet or companion that men could use with their families, and Androbot sold its Topo and BoB robots using similar strategies. The robotics company Hubotics, although the least successful in capturing the imagination or dollars of robot enthusiasts, even framed its Hubot robot more as an assistant to women's work. Since most robots seemed to be purchased and used by men, though, even this promotional strategy was likely intended as a promise that Hubot could take some household labor off dad's hands.

This uncertainty over how robots fit into family life was related to ongoing debates over what constituted an ideal participant fatherhood and men's challenges balancing work obligations, personal desires and interests, and duties to family. Robots managed these tensions in different ways. In the most limited sense, training robots could support fatherhood simply by making users more qualified for career mobility or advancement in an information economy. Pursuing these skills with a robot like Hero 1, say, could help fathers advance their careers while spending more time at home. But just being home in proximity to family was not enough to meet increasing expectations for successful fatherhood.

Here, too, some robot companies showed how their robots, like Hero Jr. or Topo, could help. Men were shown how they could use their robots with children and family or program robots to perform acts of care for other family members. In this manner, robots might help men extend the capacity of their time at home. Time spent pursuing programming, whether for career advancement or out of a personal interest, could be made to serve paternal duties. Of course, managing this balance between technological interest and domestic responsibility was not easy, and the time-intensive labor of programming threatened to clash with the time-intensive labor of childcare. Nonetheless, despite their different approaches, the promotional strategies used by Heath, Androbot, and Hubot offered models for how computers could support the role of fatherhood in the companionate family.

The Hero 1 robot produced by Heath was one of the best-selling personal robots in the 1980s. This early robot was not yet marketed as an adjunct to participant fatherhood. Rather, it was framed as a teaching tool to help men learn about programming and robotics from home, but still largely apart from domestic family life. The Hero 1 debuted in December 1982 after three years of development. Heath reported that ten thousand units had been sold by 1985, an unusually high number for the time.[33] Reporting on Heath's robot attributed the volume of sales to the company's positioning of Hero 1 as an educational tool aimed at schools and industry while also selling to a supplemental market of hobbyists at home.[34] Personal consumer use was seen as ancillary.

The Hero 1 could be considered a support to companionate fatherhood insofar as it allowed men to pursue work within the confines of the home. Advertisements for Hero 1 described it as "the world's first sophisticated teaching robot" and promised that the robot provided "hands-on knowledge of industrial electronics, mechanics, computer theory and programming as applied to robots by putting them into action."[35] The Hero 1 could augment men's programming and robotics skills to prepare them for technology careers, thereby increasing their ability to contribute to the family's financial well-being at a time of middle-class precarity. If the Hero 1 could fulfill its promise to make men more competitive on the job market, then the robot could be considered a support to his family. But Heath did not make any claims about it bolstering the emotional aspects of this relationship or encouraging time spent with wives and children.

When advertising Hero 1, Heath did not at first emphasize that the time and labor required to learn robotics and programming would be integrated into time with family. Even for those who purchased Hero 1 for use at home, Heath presented the robot as a training tool to help encourage technical knowledge more than as a domestic device. Heath is the only company discussed in this chapter that provided the option to buy its robot as a kit rather than fully assembled. This was in keeping with Heath's history as a manufacturer of electronic kits, called Heathkits, that allowed consumers to assemble their own versions of a range of consumer electronics. Since the 1940s, Heath sold their Heathkits for products like ham radios, televisions, garage door openers, home security systems, and personal computers. The kit price for Hero 1 was $1,500, compared to $2,500 for a fully assembled robot. Approximately seventy hours of labor was required to assemble the Hero 1 in kit form. But as with other Heathkits, such assembly work was often a main motivator for purchasing the product. The skills learned while assembling the robot were framed as integral to gaining the deep knowledge of robotics the kit was intended to teach. Assembling the Hero 1 before even programming it, though, had the potential to distract men from domestic duties even when using the kit at home.

By providing a robotics trainer that individuals could purchase for personal use, Heath was intentionally connecting their robot to contemporary discourses about industrial automation. When the Hero 1 was unveiled, Heath's president suggested that the robot would address "the lack of a comprehensive, affordable robotics educational system [that] has made it difficult for people to obtain the skills they need to move into robotics."[36] In fact, the Hero 1 was also marketed with an accompanying robotics "course" consisting of a 1,200-page self-study guide, which could be purchased for $99.95. Heath even offered a certificate of completion for users who submitted their final exams from the course booklet.

The robotics education that Heath's president references was a topic often raised in newspapers, magazines, and government hearings in response to anxieties about technological unemployment resulting from the uptake of industrial robotics. Experts attempting to assuage fears about unemployment suggested that any workers replaced by robots would not be laid off; rather, they would be retrained to operate, maintain, or program industrial robots.[37] Often it was blue-collar workers who were being laid off in favor of industrial robots. But Heath's marketing depicted white middle-class men using Hero 1, and Heath's robots targeted those who could afford a $1,500 hobby. It was these men who could purchase the tools to learn about robotics and computer programming, skills deemed to be less easily replaceable by machines and therefore necessary training in a changing economy.

While not yet addressing the expectations of participant fatherhood, the Hero 1 could help men in their economic role in the family by reinforcing the value of their labor. As a robot designed to teach programming and robotics at home, Hero 1 was a part of larger contestations over the value of men's work in the 1980s. Neda Atanasoski and Kalindi Vora discuss the long history through which the robot has stood in for attempts to classify and distinguish the value of different kinds of labor. As they argue, "Since the first industrial revolution, automation has signaled the threat of the replaceability of specific types of human functions and human workers that are racialized and gendered. This is because the tasks deemed automatable, including manual labor, blue[-]collar factory work, and reproductive and care work, were regarded as unskilled and noncreative—work that could be done by the poor, the uneducated, the colonized, and women."[38] Heath provided a privatized solution to larger anxieties about deindustrialization and unemployment: robotics training that could be practiced in the middle-class home through the purchase of a consumer robot. Heath's marketing seemed to target a professional class of managers, then, for whom computing and robotics were less of an existential threat and more a path to economic advancement. By buying these robots to privately develop robotics skills, hobbyists could further distinguish their skills and labor from the blue-collar work vulnerable to replacement by industrial automation.

Reporting on robots in industry often pointed to the career opportunities for those who programmed them, despite concerns that these machines would replace other types of labor deemed more repetitive and lower skilled.[39] In fact, a science reporter for the *New York Times* even seemed to map concerns about outsourcing and skilled labor onto the origin of the chips used to make the Hero 1. Discussing his experience assembling his Hero 1 kit, the reviewer describes how the chips "seemed to carry labels from everywhere except the United States. Malaysia. Indonesia. Korea.

Even war-torn El Salvador. (Where will the plants be built when the ulti-mate inexpensive labor force, robots, take on the chip-making job?)"[40] He notes his relief at seeing that Hero 1's "brain," or central processing unit, was designed in the United States. These distinctions helped to mark some types of labor, that of white American middle-class men and knowl-edge workers, as more valuable and difficult to replace by automation or outsourcing. Men learning robotics and programming from their Hero 1 were assured of the continued value of their labor as they trained on these "smart" skills at home.

In advertisements, the Hero 1 was usually shown on a work desk with a lone man tinkering with the robot or reading the accompanying docu-mentation. In one typical example, a man works on the Hero 1, which is shown without its plastic shell, exposing the electronics inside (Figure 6). His workspace is sparsely inhabited except for robot parts and tools. The imagery in Hero 1 ads refigured the home as a robotics workshop, sug-gesting that those who spent enough time with their robot could gain necessary skills and mastery over technology. These promotional images of the robot did not specify a domestic space where it might fit and made no promises about its potential as a domestic servant, companion, or inte-gral part of family life. In fact, other members of the household did not appear in advertisements for the Hero 1.[41] This is an image of robotics work not yet part of familial relationships and routine.

By offering personalized robotics training in the comfort of one's own home but in a space apart from family life, Hero 1 could enable what historian Steven Gelber refers to as domestic masculinity, not to be con-fused with masculine domesticity. Initially theorized to describe DIY culture in the early to mid-twentieth century, domestic masculinity, Gel-ber argues, was a role "that suburban men created so that they could actively participate in family activities while retaining a distinct mascu-line style."[42] Whereas masculine domesticity referred to men's engage-ment with domestic activities and childcare, domestic masculinity allowed men to claim part of the home as their own and use this space to carry out tasks traditionally coded as masculine. Heath's Hero 1 could bring the masculine-coded computing and robotics hobby into the home; conse-quently, in advertising imagery, this work appeared to be set apart in a masculine workspace distanced from domestic routine.

Heath envisioned Hero 1 as an instructional robot that would build programming skills. But when the robot did enter the home, user accounts suggest it was not easy to separate men's work in a robotics workshop in

The text within the advertisement image reads:

Move into the world of robotics with HERO 1

① **HERO 1 is the ideal robotics training tool**

HERO 1 is a completely equipped trainer designed to demonstrate every major robotic concept and all the basic systems found on the modern and increasingly sophisticated robots and automated machines of industry.

Bring the concepts of high-technology robotics to life with practical hands-on training using the teaching robot, HERO 1. This computer-controlled, electro-mechanical device allows you to explore and work with all the fundamental components and circuitry associated with robot technology. Completely self-contained, HERO 1 is capable of interacting with you and its environment. It detects light, sound, motion and obstacles in its path; and it can travel over a predetermined course. When using its optional arm, the robot can be programmed to pick up small objects with its manipulator. And with its optional voice synthesizer, HERO 1 can even speak in complete sentences.

An intelligent robot, HERO 1 has a computer brain consisting of an on-board 8-bit 6808 microprocessor. Following programmed instructions, the microprocessor can guide HERO 1 through complex maneuvers, activate and monitor sensors, and modify the robot's actions as a result of sensor or real-time clock inputs. Memory consists of 8K of ROM and 4K bytes of RAM. This can be increased up to 56K with the addition of an optional memory expansion board. Preprogrammed ROMs are also available for installation on this board that allow HERO 1 to demonstrate its many capabilities.

Program HERO 1 using any one of several methods. From the top-mounted 17-key keyboard you can easily enter, verify and modify programs as well as access any of the seven operating modes. Also access any of the microprocessor's registers through the keyboard. An attachable Teaching Pendant lets you manually control all motor and arm movements or store them in memory for later duplication. A rear-panel serial port allows programs stored in memory to be transferred to a cassette tape for later reloading and use.

Two totally new methods to control HERO 1 are now available. One is a remote, radio frequency-controlled transmitter available in 3 models, each operating at a different frequency. It controls all keyboard and Teaching Pendant operations from up to 100 feet away. An RS-232 connector on this accessory also permits a computer to remotely operate the robot. The Remote has a self-contained re-chargeable battery that provides operating power for up to five hours. Control is also provided through an optional RS-232 Interface that plugs onto the top breadboard providing a direct link between a host computer and HERO 1.

Light, sound and motion detectors plus a sonar ranging system gives HERO 1 the ability to see and hear. The sound detector hears over a 300 to 5000 Hz frequency range while the light detector sees over the entire visible spectrum and into the infrared spectrum. The motion detector senses movement up to a distance of 15 feet and the sonar system determines the range between objects and the robot.

HERO 1 is completely equipped to demonstrate every major robotic concept

Hexadecimal keypad and LED display

Teaching Pendant controller

NEW optional RF Remote Control Accessory enables direct and computer control of HERO 1 from up to 100 feet away

Figure 6. An advertisement for Heath's Hero 1 depicts the user in a sparse, workshop-like space. This image was included in the Heath company's Christmas 1984 Heathkit catalog.

the home from their embedded family relationships. A review of Hero 1 in *Popular Mechanics,* for example, opens with the author's daughter asking plaintively when Hero would be back home in response to the robot's absence while being photographed for the magazine.[43] Similarly, in an article in *Popular Computing,* an English professor detailed the time he spent using Hero 1 with his family, describing how he programmed it to greet his children when they got home from school and to play a *Star Wars*–inspired game with his son.[44] Additionally, users could purchase third-party programs that enabled the robot to play games or act like a pet. These examples show that the Hero 1 took up domestic roles, like entertaining the family, that went beyond functioning as a training tool for the isolated home workshop.

The potential for integrating robots into family dynamics and extending paternal attention was taken up more explicitly two years later, in 1984, when Heath introduced Hero Jr. Whereas the Hero 1 was primarily a teaching tool, Heath promoted Hero Jr. as an integral part of the family. When development on the Hero 1 began in 1979, the nascent market for home computers was still strongly associated with hobbyist interests. The marketing of computers to a wider domestic audience had considerably expanded by the time Heath launched Hero Jr. This was even reflected in the fact that Hero Jr. was produced by Heath's consumer products division, unlike the original Hero 1. Hero Jr. also sold for a third of the price of Hero 1, indicating that it might be bought more as a playful technology than a serious robotics trainer. Still, Heath tried to strike a balance between positioning it as a tool for learning robotics and an entertaining family technology. The Hero Jr. owner's guide explained that the robot was not only a companion but also able to perform useful functions "like any well trained pet."[45] Heath informed potential consumers that no prerequisite computer programming skills were necessary to get Hero Jr. working in their homes, but also promoted the option to design one's own programs.[46]

Advertisements for Hero Jr. emphasized the robot's fit with domestic life, helping bring men's work with robots closer to the family. Unlike the images of isolated men in undefined workshop spaces used to sell Hero 1, Hero Jr. ads placed the robot within the family circle. One Hero Jr. ad shows a Christmas morning vignette, in which a young child warmly embraces the robot as parents and sibling look on (Figure 7). Another image shows a whole family gathered around the robot with the owner's manual in the mother's lap as they all turn their attention to the device.[47] When Hero Jr. was released in 1984, Heath described it as "more like a household pet than just a mechanical device."[48] Rather than show the robot in a professional or workplace setting, these images presented Hero Jr. as a technology for family togetherness and even as a gift that would be beloved by everyone in the household.

Although Hero Jr. was a family robot, its promotional materials suggested it was targeted to men who might share their interest in Hero Jr. and the pleasure of robotics with their families.[49] The robot came with programs to sing songs like "Daisy" and "America the Beautiful," recite poetry, carry on small talk, and even lead the family in an exercise routine.

Figure 7. An advertisement for Heath's Hero Jr. embeds the robot in the family. This image was included in the Heath company's Christmas 1984 Heathkit catalog.

Heath released additional cartridges that could quiz children on math, sing nursery rhymes, and do party tricks. With such family-oriented programs, Hero Jr. could support men in achieving the expectations of participant fatherhood without sacrificing their computer hobby. Time spent on robotics was also time spent together with family.

Yet, in some instances these applications seemed to promise that robots would carry out acts of care on their users' behalf, rather than encourage fathers to be more engaged with family life. For example, one of the first things Hero Jr. could be programmed to do is to remember and announce special occasions. The examples offered in the owner's manual included "a child's birthday, an anniversary, or other regular event."[50] Men did not need to remember an anniversary or a child's birthday because they could program Hero Jr. to announce such occasions or to carry out preprogrammed sequences to celebrate them. An advertisement for the robot promises, "You'll never be embarrassed again by missing that important date."[51] Thanks to this feature, the user could spend time learning to code and satisfying his curiosity about robotics while producing a program that would perform familial care in his place. By providing a tool that could be programmed for familial or childcare uses, the robot helped users juggle professional development and child-rearing but potentially stood in for his undivided attention.

Androbot was another company selling robots in the 1980s as a tool to extend participant fatherhood. Androbot was started by Nolan Bushnell, famous for founding Atari and, through sales of the Atari VCS, helping bring the video game medium into millions of American homes. In 1983, Androbot sold a robot called Topo, which differed from the other robots discussed in this chapter because it was not fitted with a microprocessor. Topo had to be programmed using an Apple II computer and operated remotely by signals from an attached transmitter. In an Androbot newsletter, Topo was discussed both as a robot and as "a mobile extension of your personal computer." Androbot also marketed another robot called BoB (Brains on Board) that would have had a complete microprocessor-based system. BoB never made it to market because Androbot went out of business before the robot was available to sell. But this robot inspired just as much discourse as Topo and was frequently promoted and demonstrated at robot and computer fairs. Androbot described their robots as friendly and integrated into domestic life: "We have conceived our Androbots for home use—to interact socially with people. You will discover that Androbots have personalities and capabilities to make them entertaining, safe, and educational." As the same newsletter summarized it, an Androbot robot was "not an appliance, but a friend."[52]

Like Heath did with Hero Jr., Androbot also advanced the idea that their robots would function as a way for men to extend their capacity to function as an integral part of the family. One promotional image, for example, shows a father who has gathered his children around the computer to program Topo.[53] In this image, he is able to maintain the attention and interest of three children while also spending time on the computer. Androbot often made claims about their robot's assistance to fathers by projecting situations in which the robot could act on behalf of the user if he were busy or otherwise occupied. The company was liberal in speculating on capabilities its robots might have in the future. Androbot spokespeople and promotional materials laid out situations in which Androbot robots would extend their owner's mobility and management of domestic tasks: in the future their robots could greet guests at parties while the hosts are in the kitchen, or follow kids around to teach them different topics. In an interview, Androbot president Tom Frisina even laid out a scenario in which programmers could teach BoB to greet children and babysit while parents were still at work—a feat far beyond the robot's technological capability.[54] These prognostications suggest that Androbot was presenting their robot to eventually serve as an extension of paternal care. Here the promise is not necessarily a robot to use with the family, but one that could stand in for fathers, helping them juggle their responsibilities at home and work.

Heath and Androbot largely targeted their promotions toward men who might use the robot to pursue their hobby while also carrying out technologically assisted family- and childcare tasks. But on some occasions, personal robot advertising tried to depict how these robots might interact with women and serve their activities in the home. Examples in which personal robots aid women in their domestic labor may not obviously appear to relate to fathers' roles as part of companionate family life. But it is likely that these images of robots as domestic helpers were intended to appeal to men by promising to take over some of the domestic labor they were increasingly being expected to share with women. Even if men did not necessarily see robots as their substitutes in aiding women with domestic labor, this promotional address might help men convince wives of the robots' usefulness to the everyday life of the home.

Androbot advertising occasionally showed women at home interacting with Topo and BoB. An image demonstrating the Topovac, a never-released vacuum attachment for the Topo robot, depicted a woman reclining on a couch reading a magazine while the device vacuums a huge spill on the carpet in the foreground.[55] However, this type of image, in which the robot replaced women in their domestic labor and freed them to enjoy their leisure time, was used infrequently. A more common strategy showed personal robots assisting women as they multitasked around the house. A pamphlet advertising BoB showed a woman kneeling beside the robot as she placed a bottle of wine in its Androwagon, a plastic carrier attachment. The accompanying text describes that in the future BoB would be able to help at home by greeting party guests.[56]

Another robot company, Hubotics, represented women more centrally as the operators or benefactors of domestic robots. Hubot debuted at the Consumer Electronics Show in January 1984 and retailed for $3,495, making it one of the most expensive robots sold commercially at the time. It is possible the price was elevated because Hubot came equipped not only with a computer and keyboard, but also a floppy disk drive, twelve-inch television screen, Atari video game console with two joysticks, AM/FM radio, and stereo cassette player. Saddled with all this technology, Hubot cut a more imposing figure than other robots. Its components were stacked in a tower that was nearly four feet tall and weighed almost 120 pounds. In terms of sales, Hubot was also the least successful of all the commercial robots. It was reported Hubotics only sold fifty-five robots in the first year of operation before quickly dissolving.[57]

Advertising for Hubot connected the robot closely to domestic tasks. Hubot was described not only as a robot but as "the ultimate intelligent home appliance."[58] This label as an appliance associated it more closely to previous kitchen or household technologies that targeted homemakers. Additionally, Hubotics decided to sell this personal robot in department stores and catalogs, bypassing computer and electronics stores that reached more specialized, male computer hobbyists. According to an advertising industry publication, the New York department store Abraham & Straus ran television commercials for Hubot that introduced it as "Mother's Little Helper" and "portrayed Hubot as a kind of live-in maid who 'walks, talks and serves breakfast in bed.'"[59] Hubot also diverged from other robots because of its pared-down, user-friendly interface. Many of Hubot's features did not require even simple programming skills, but functioned via menu screens. This potentially made the robot more accessible to users who were not interested in or knowledgeable about programming.

Promotional images circulated by Hubotics pictured the robot as helper to women doing tasks in the home. In one such image, a woman cuts vegetables on her counter while watching the TV screen embedded at the top of Hubot's body.[60] Hubot is staged with its own cutting board holding a partially sliced baguette, as if it is also in the middle of food prep. The device is presented as a work companion, helping the woman as she does her cooking, even though Hubot's usefulness in this situation is limited to serving as a tray and mobile television screen.

The images of the vacuuming Topo and meal-preparing Hubot rely on well-worn conventions for selling electronic appliances, indicating that robot makers cared about how their products appeared to women, even if they knew that men were most likely to buy them. Scholars have noted that since the 1920s, electronic domestic appliances have been sold with promises to liberate housewives from domestic drudgery and allow them to freely engage in leisure activities.[61] Describing the robot featured in GM's educational film *Leave It to Roll-Oh* (1940), for example, Lynn Spigel argues that "the drudgery, loneliness, and submission of women was transformed into play, companionship, and dominion through the wondrous technology of . . . 'thinking things.'"[62] Images of Topo and Hubot helping women at home appear very similar to fantasies like these from decades earlier.

Yet, Hubot advertisements differed from the longer history of electronic appliance marketing because they also addressed women who were presumably balancing a career outside the home with their domestic duties. A promotional image of the Hubot depicts the robot helping such a woman. The ad opens with Hubot addressing its owner: "OK, Gail. Call you at 10." In the background, Gail appears to have just arrived at home and directed the Hubot to set an alarm. She is still dressed from her day outside and her high heels have been removed and placed haphazardly on the nearest surface—in this case, on top of her grand piano. Gail seems to be an upscale consumer and career woman. The accompanying text focuses on Hubot's ability to assist in housework, to serve as a companion "whenever you are at home," and to entertain and educate the family.[63] As more women in the 1980s were balancing participation in the paid labor force with duties at home, a robot like Hubot, it was suggested, could potentially pick up some of this extra work.

Marketing and promotions for personal robots from Heath, Androbot, and Hubot used a variety of strategies to situate them as domestic technologies that could aid men, and occasionally women, in satisfying their responsibilities to children and family. At times this was achieved by allowing men to train in robotics skills from home, helping them secure their earning abilities to support the family. Occasionally, promotions suggested that robots might compensate for absent men by helping women in their domestic duties. But more ambitiously, robots were presented as tools to support and extend men's capacities in fulfilling the expectations for participant fatherhood and relate more closely to their families through a shared interest in computing—using robots to care for children, show them attention, and initiate them into shared technical familiarity as well.

Reviews and discussions of robot use show how individuals and families incorporated the ideals proposed in advertising to their own experiences navigating companionate relations with robots at home. Some robot owners described the many ways they successfully incorporated robots into their family routine. Fred D'Ignazio, a frequent contributor to computer magazines in the 1980s, offered anecdotes about programming Topo to wake his children in the morning and to entertain them in various ways. He even describes the attempts his children made to program Topo to do small tasks.[64] A *New York Times* article on home robots reported on one family that housed a Hubot, bought by a computer software developer father. The story explains that although his purchase was initially a source of disagreement—his wife had wanted to use the money to buy a fur coat—this was resolved once she saw that their two-year-old son took to the machine. The article recounts that the child named the robot Bumpky, followed it around everywhere, and learned the alphabet from the robot's instruction. The father said he used his big-ticket purchase to develop his programming skills and serve as a companion to his son.[65]

There is less evidence that robots were successfully marketed to women as assistants to their domestic tasks at home. The same article above that discussed Hubot also mentioned a woman who bought a different home robot, the RB5X, and used it with children in her child development career. In *Family Computing* magazine, a woman reported on how successfully her family integrated Topo into their companionate relationships. The author describes programming Topo with her family and how her husband used it as an opportunity to tell the family about the future robots from his employer, General Electric.[66] She recounts how the robot helped bring the family together, encouraging them to talk and share in ways she felt were increasingly less common. Also, although she does not entirely specify her work situation, this woman describes how she used Topo to teach her adult education computing class. In this instance, the robot helps the woman with her work outside the home, but she also describes it as a technology that facilitates her husband as he bonds with the family about his work.

These accounts show how some families attempted to use robots to help men achieve ideals of participant fatherhood in ways that resembled advertising promises. As historians of media and technology have argued, promotions for domestic technologies—ranging from early twentieth-century electric kitchen appliances to early twenty-first-century social robots and networked smart home devices—have often been marketed in ways that suggest they can strengthen the companionate family. Historically, these promotions often showed women liberated from domestic tasks through technology, while also promising to maintain conservative ideas about self-sufficient autonomous nuclear families and traditional gender roles. This argument has been invoked to describe robots and robot representations over a large historical span, from the aforementioned fictional Roll-Oh domestic robot of GM's sponsored film to contemporary smart home technologies.[67] Discussing a 2014 promotional video for the Jibo home robot, for example, Atanasoski and Vora argue that advertising for these robots "inserts them into the home in a way that normalizes the white U.S. middle-class nuclear family form. More specifically, it advances the norm of the household as an autonomous and self-sufficient economic unit, even as it proves that unit to require substantial and invisible support labor."[68]

Like these other technologies, home robots in the 1980s were marketed as tools to help sustain companionate family relations and to present the middle-class family as an autonomous, self-sufficient unit. Yet, they did so primarily by speaking to men, addressing concerns about how they could integrate their interest in programming and robotics with the increased pressure to spend time with their children.

Relating with Robots: Cute Robots and Nurturing Masculinity

Robot marketing suggested that these devices could help men to be good fathers by programming them to complete caretaking tasks or by including the family in robot interactions. Yet, even if users were able to incorporate robots into their family relations in this way, the ideal of participant fatherhood was not limited to spending time with one's children—there were also increasing expectations for men to be caring and nurturing. In the 1980s, a common lament was the difficulty of embodying the necessary affectionate masculinity that came to be associated with the new style of participant fatherhood. Men who felt that their fathers had not provided a role model for this style of parenting could attend workshops or find published guides that described other men's experiences grappling with these new expectations.[69]

Could personal robots also be a resource in fostering this nurturing style of affectionate parenting? In fact, robot manufacturers often framed robots as pets or children that men could take care of and grow intimate with through their programming. Companies even described their robots as cute and discussed the impact of cuteness on design decisions.[70] These comparisons to pets and children and discussions of cuteness were likely motivated by the need to compensate for the robots' limited power and functionality. Nonetheless, in trying to justify these products, those making and using them developed a framework to make sense of robot interaction and programming as a kind of caring activity. In this regard, robots could help men achieve the nurturing requirements of participant fatherhood not by providing a role model of an attentive father to emulate, as workshops or fatherhood classes might, but rather by encouraging men to practice a nurturing and preservative care for their robots. This was quite a different model of computing from the more domineering versions of masculinity that had come to be associated with professional, hobbyist, and even much general interest computing. There is no indication that robot manufacturers purposefully created robots to teach nurturing to men; however, this mode of robot activity hinted that learning to program could be related to nurturing, and thus aligned with shifting ideals for masculine and paternal affection.

Robot cuteness was expressed in part through a loose anthropomorphism. The Hero 1 and Hero Jr. were both about twenty inches tall and about sixteen inches wide with a roughly cuboid shape. Both robots also had a small cuboid "head" at the top of their bodies where many of their sensors were located. The transmitter and receiver of Hero 1's sonar were arranged to look like two eyes in the robot's face. On the Hero 1, this feature could rotate 320 degrees, which better simulated a head turning to scan its surroundings. This is also where the optional arm and gripper could be attached to the robot. Overall, both Heathkit robots gave the impression of squat and diminutive bodies, which contributed to their cute design.

Androbot's robots were also embodied in a form that encouraged users to view them as friendly and familiar objects of affection. Topo and BoB were taller than the Heathkit robots, at thirty-six inches, but still similar in size to a young child. Even more than the Hero Jr., the Androbot robots were designed to look like anthropomorphized and friendly companions—or as Androbot's president describes it, "a playful companion but one that looks somewhat futuristic."[71] Topo's and BoB's anthropomorphic features were more defined than those of the Heathkit robots. Both robots had a round head marked with two round eyes and a triangle-shaped mouth. In the Topo, these features were entirely cosmetic because this robot did not have any kind of sensor input. BoB's eyes were made up of ultrasonic and infrared sensors; unlike Topo, BoB's head could turn left and right to allow the robot's sensors to scan the room. Below their round heads, each Androbot robot had broad shoulders and a somewhat defined waist before widening to the area where the two-wheel system was located. One robot reviewer described Topo as looking like a little snowman.[72] These anthropomorphic features did not have a clear technological function but added to the relationship Androbot was trying to create. As a company press release explained, "From the very look and shape of the head and body style, Topo tells you that he's going to be a cheerful, loyal friend."[73]

Contributing to the impression that these robots were to be treated affectionately like pets or children, they were not designed to be conspicuously gendered or hypersexualized. Rather, there were some indications that these robots were meant to be read as exhibiting a juvenile masculinity. Promotional material and news coverage tended to refer to personal robots as "he" or "it," rarely using feminine pronouns. Robots were given names coded as masculine like BoB and Hero Jr.[74] Heathkit, Hubotics, and Androbot robots were equipped with low-pitched speaking tones, a quality most associated with men's voices. Even here, though, Androbot and Heath provided users with instructions to adjust pitch and volume and therefore alter the presumed gendering. Beyond these indications of a boyish masculinity, the gendering of robots was ambiguous: they were not hypersexualized fembots or hypermasculine machines. In this respect, personal robots diverged from representations in science fiction and fantasy like the docile androids of *The Stepford Wives* novel (Ira Levin, 1972) and film (dir. Bryan Forbes, 1975) or the animatronic cowboys of *Westworld* (dir. Michael Crichton, 1973).[75] Instead, they were overwhelmingly presented as cute and childlike objects to inspire preservative care and companionship.

Many of these features of personal robot design were consistent with aesthetics of machine cuteness. As theorized by scholars like Anthony P. McIntyre, "machine cute" is characterized by indicators of vulnerability, lack of bodily integrity, limited linguistic capacity, and cognitive neoteny.[76] These were all features shared by Heath, Androbot, and Hubot robots: their diminutive size (ranging from two to under four feet tall), rounded bodies and softened edges, slow motion, limited speech, and tendency to collide with obstacles. McIntyre argues that machine cuteness facilitates robots' pedagogical role in alleviating fears of automated labor.[77] Specifically in the 1980s, when fears of technological unemployment were high, cute personal robots likely contributed to these devices' capacity to make automation seem more familiar and perhaps even more friendly. Articles about personal robots often referred to their deficiencies in completing practical tasks to jokingly assure readers that they would not be replaced by automation anytime soon.[78]

But machine cuteness also functioned to shape the relationships that would develop between robot users and their new robot companions, encouraging a kind of affectionate and caring form of relationship. In more recent years, machine cuteness has taken on a significant role in the design of sociable robots. Jennifer Rhee has discussed, for example, how MIT's Cynthia Breazeal designed sociable robots like Kismet to "evoke combinations of machine, animal, and human infant in order to encourage humans to take the lead in familiar adult–child, teacher–student, or parent–child dynamics. . . . And as humans interact with Kismet, they also learn to become increasingly deft at reading Kismet's emotions and making their own emotional expressions legible to Kismet."[79] More recent sociable robots like Jibo have been called companion robots, and researchers have discussed their ability to maintain durable relationships with users through cuteness.[80] At the time of writing this chapter, delivery robots have also started to appear on U.S. city streets. Both the designers of these robots and social commentators highlight them as cute and adorable, presumably to deter rough or violent reactions from passersby.[81] But even back in the 1980s robots were designed to be cute to encourage certain types of caring relationality. At a presentation for the Boston Computer Society in 1983, for example, Androbot engineer Bill La spoke to the audience about the design tradeoffs required to effectively present Topo and BoB as cute technologies that could be accommodated to the home.[82]

Theorizations of the cute discuss it not only as an aesthetic, but an affective and emotional response elicited by cute objects. In other words, cuteness defines a relationship as well as an aesthetic. Discussing verbal outbursts in response to cuteness, Sianne Ngai theorizes, "Cuteness cutifies the language of the aesthetic response it compels, a verbal mimesis underscoring the judging subject's empathetic desire to reduce the distance between herself and the object."[83] Similarly, Lori Merish argues that the "culturally sanctioned response to the 'cute'" is a "protective cherishing" and "preservative love." Cuteness is based on an imbalance of power between subject and object where the powerlessness of the object is aestheticized and appeals to the user for care: "The cute *demands* a maternal response and interpellates its viewers/consumers as 'maternal.'"[84] The cuteness of personal robots encouraged male hobbyists to interact with their childlike robot with this kind of parental care. At a time when men were expected to be more affectionate and caring but did not necessarily have the skills or time to do so, robots were tools to help train them while still satisfying their technological interests.

In addition to designing cuteness into home robots, these companies modeled a relationship of preservative care for their robots in advertisements and publicity photos. Promotional images indicate how the robot's small body required certain postures of diminution from its user to interact with it. Not only were Hero 1 and Hero Jr. diminutive robots, but their size required users to make themselves small by crouching, kneeling, or sitting as they directed their attention to or manipulated the robots. A two-page ad that ran in many computer magazines featured Nolan Bushnell, Androbot's founder, sitting cross-legged next to Topo with his arm slung around the robot's shoulder (Figure 8). Sitting in this pose Bushnell, despite his suit, is made to look more playful and affectionate. This paternal pose was a common one; other members of the Androbot team were often pictured in similar images with their arms around Topo or BoB, including president Tom Frisina and engineer Doug Jones.[85] The wide shoulder area of this robot even conveniently served as a resting place for this kind of embrace.

Figure 8. Advertisement for Androbot's Topo robot. Androbot founder Nolan Bushnell is shown in an affectionate pose with the robot. The spot ran in the December 26, 1983 / January 2, 1984, issue of *InfoWorld*.

Hero Jr.'s small size was highlighted by many of the promotional vignettes in which it was placed as well. One advertisement shows a child of approximately seven wrapping her arms around the robot with her head lying on Hero Jr.'s head. This pose does not only highlight that the robot is meant to be an object of affection, but also accentuates its similarity to the child. In another Heathkit advertisement, a family of four gathers around Hero Jr. and all but the seated mother are forced to kneel or crouch to examine the robot closely (Figure 9). In the postures users take to engage with personal robots, we can see another way, beyond the verbal, in which

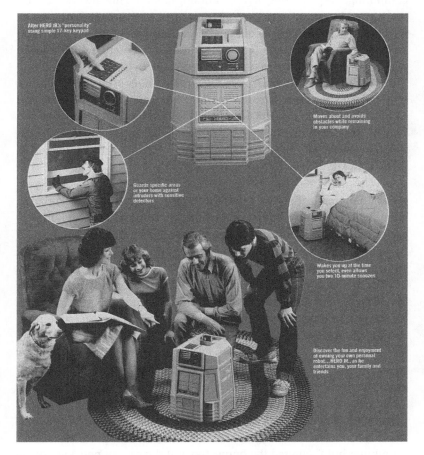

Figure 9. An advertisement for Heath's Hero Jr. shows users crouching and kneeling to engage with the robot. This image was included in the Heath company's Christmas 1984 Heathkit catalog.

"cuteness cutifies" the subject engaging with a cute object. Users bend to close the physical distance between themselves and their robot companions, and in doing so they take on similar qualities of diminution.

A relationship to robots of paternal care for an embodied domestic object was an alternative model to conceptions of programming that were dominant at the time. Computer historians have charted the shifting gendered associations with computer labor since the 1940s. Although there are currently fewer women in computer science education and programming careers and strong cultural associations between masculinity and computing, the gendering of programming labor did not always function this way. Before there were even physical technologies called computers, women doing ballistics calculations were called "computers." Jen Light discusses the way the term "computer" shifted to describe the hardware and the women programming computers came to be called "operators." The work of operators, which we would now think of as programmers, required advanced mathematical training but was categorized as clerical work. This work was thought of as repetitive, mechanical, and menial, the type of labor male engineers felt was beneath them. As Light argues, in the 1940s "what we now call programming . . . fit easily with notions about women's work."[86]

It was in the 1960s and 1970s, well before the wide availability of personal computing, that this gendering of the programming occupation shifted. Nathan Ensmenger explains that as demand for programmers grew and compensation for the job increased, men moved into the field in greater numbers and worked to combat the occupation's feminized associations. Even though the actual labor of programming did not necessarily change, professional authority for programmers was developed by distancing programming skills from clerical work and everyday routines. An image developed that set programming in direct contrast to interpersonal skills and concerns for the routine labor of self-care, embodiment, and everyday life.[87] The talent of the programmer could be seen in how he neglected interactions with others, ties to family or domestic duties, and even the need to shower or cut his hair.

In many instances, when personal computers entered the home and became more widely available in the 1980s, this association of programming with the masculinity and compulsive passion of the "hacker" or "computer bum" entered the home, too. But the case of personal robots suggests that the intersection between computers, domesticity, and companionate family life also offered challenges to this decontextualized and disembodied version of programming. Personal robots proposed a different association with programming in which developing computer labor also meant taking care of a robot or mobile computer and tending to its interactions in the space of the home. This modeled a version of paternal and nurturing care idealized in contemporaneous discourses about fatherhood.

In addition to their cute embodiments, other features combined to encourage this mode of paternal care by contributing to the impression that robots needed protection. The cute aesthetic design of personal robots helped guide the response to their frequent dependency and powerlessness. These robots in fact were extremely constrained in their functions. All the robots discussed in this chapter could move in multiple directions on their wheels, but not quickly. Hero Jr. and Androbot robots could not physically manipulate their environment, lacking robot arms or grippers that would allow them to lift objects or pour drinks. Hero 1 came equipped with an optional arm and gripper, but it could carry only up to one pound. The arm could move in a few directions, but it did so very slowly.

Many of the robots came with synthetic speech capabilities, but this meant that programming them to speak often required laborious inputs to instruct them phoneme by phoneme. Additionally, most of these robots had only limited sensors that provided them information about their environments. Hero Jr. could respond to bursts of sound and, for example, count the number of times a user clapped. Heath robots had light sensors and ultrasonic sonar and motion detectors intended to help the robots avoid collisions. Still, these were not robust systems. Many users found robot sensors deficient and reported that units were prone to crash into walls or other obstacles.[88]

Home robots were vulnerable to accidental damage and constrained in their functionality. This contributed to a mode of interaction in which users were primed to expend preservative care and attention over the robot. Androbot's Topo robot represented the most extreme version of robot as vulnerable figure. Unlike the other personal robots like Hero 1, Hero Jr., and Hubot, all of which came equipped with microprocessors that could store programs and direct their movements, Topo was not an autonomous device. As mentioned above, it could only be programmed with the use of an Apple II via an attached radio or infrared transmitter supplied with purchase.[89] With this setup, some complained about the way that walls and other features of domestic architecture could impede the communication between robot and computer.

The process of getting Topo to walk highlights how robot interactions required care. The manual informed eager users that they could immediately see Topo in action before learning any of the detailed programming commands to control it. The robot could be directed to undertake some limited movements by way of four buttons on top of its head. To guide Topo the user had to closely follow the robot to stay in contact with its head switch at all times. That way, the user could turn Topo when needed and stop the robot before it collided with objects in the room. The robot's physical integrity depended on the user monitoring it in this way. Topo had no ability to avoid collisions on its own because it had no sensors or awareness of its environment. In guiding Topo this way, the user was forced to walk slowly (the robot had a maximum speed of fifty centimeters a second, or a bit more than a mile an hour), shuffling beside the robot and remaining vigilant about potential threats.

This mode of conveyance also applied to the remote control of the robot as it carried out preprogrammed sequences. The instruction manual even directs users, "Because you are just learning how to pilot Topo around the house, and are just learning what kinds of motions Topo can make, it is a very good idea for you to follow Topo around as he goes on his excursions."[90] Any fantasy that robots might facilitate the user's unhindered mobility is immediately contradicted as the user must exercise great care in keeping the robot safe. Walking with Topo resembles accompanying a toddling child more than a powerful robot.

This understanding of robotics labor as a form of childcare was mimicked in reviews and accounts of men's experience with robots at home. For example, the father mentioned above who reviewed Hero 1 for *Popular Mechanics* consistently referred to the robot as a "baby," "newborn," or "little guy" and compared his process of assembling and programming it to raising a child: "In less than two weeks of partial evenings, there he stood—a baby robot with cute little tubular eyes, tiny hand ever ready to grasp your finger, little forehead wrinkled with buttons and a pleasant, ready-to-learn attitude." The reviewer goes on to describe learning to program each aspect of the robot through the metaphor of child-rearing. Using the voice-synthesis feature is teaching Hero his "first words"; learning how to program the robot to move is "helping Hero take a few steps"; discussions of the battery are framed as wonder at "where he gets all his energy."[91] This shows how some men internalized the guidance to think of their robots as children, programming them in a way that focused on the devices' perceived vulnerability. Here robots worked as support to participant fatherhood by encouraging men to experience their activity as childcare.

Describing his family's experience with multiple robots in their home, D'Ignazio even presented robots as the prime example of a relationship to computing he referred to as computing intimacy. D'Ignazio made a case in *Compute!* magazine advocating for computer intimacy as opposed to computer literacy. Whereas computer literacy meant a user could manipulate programs and understand the inner workings of the unit, computer intimacy was about feeling comfortable living and interacting with computers without necessarily knowing how to program or tinker with them. Robots were useful, according to D'Ignazio, not for any particular household function they could perform (e.g., vacuuming), but rather for how they teach users to relate intimately with computers as friends.[92]

Some accounts suggest that users were not just encouraged to take greater care for robots; working with these technologies also made men more aware of the particular domestic context in which computing occurred. Reviews and articles on these devices noted the challenges of operating robots in domestic space and suggested technological fixes. In *Robotics Age*, one hobbyist commiserates with other robot operators, "You have probably already encountered the problem of running into warm bodies, be they dogs, cats, or people. Of course, the last thing you want is a robot that runs over the spouse or household pet."[93] The writer shares how he installed additional sensors on his Hero 1 robot to alleviate this issue and provides readers with the instructions to do the same for their own home.

Commentators and robot experts described the challenge of operating domestic robots at home by referring to structured and unstructured spaces. Structured spaces like factory floors were considered easier to automate because robots installed there would not need to adjust to changing conditions. The home, in contrast, is an unstructured space that required robots that could sense and respond to unpredictable pathways.[94] Acknowledging the home as an unstructured space may show the begrudging awareness of its complexity and the labor required to maintain it—even if the impulse driving this awareness originated from the desire to rationalize and simplify it. Here, too, robots encouraged men to become integrated into the specific patterns, routines, and conditions of the domestic space in which they were carrying out their computing work.

Of course, these engagements with personal robots did not always translate to better familial relationships or greater engagement with home life. In some cases, this form of caring for robots as children threatened to impede rather than facilitate childcare. The aforementioned English professor's account of his time with Hero 1 embodied the struggle between using programming to connect with children and the risk of pushing them away. The reviewer, Peter Owens, explains, "Having Hero around the house is a bit like parenting a robot toddler . . . studying a toddler's behavior can be enormously useful . . . as a robot toddler, Hero 1 fulfills a similar purpose." Not only does he compare Hero 1 to a toddler, he also describes using the robot as part of his own child-rearing. Owens relates how he programmed the robot to greet his children and play games with them, finding joy in their happiness and his robot's expanding usefulness. Yet, he also notes his young son's growing resentment toward him for spending too much time with the robot.[95] This account hints at the challenges of incorporating robots and programming interest with everyday child-rearing duties. While a robot can extend a father's caretaking abilities by entertaining children, that benefit must be weighed against the time and attention required to program it in the first place.

Discussing a different domestic computer application, 1980s recipe software, Maureen Ryan describes an issue with this type of technological solution. As she argues, programs like recipe software "seem expressly designed for those who treat cooking as a hobby—who can build the recipe program or tinker with food hacking projects on the weekends—rather than someone in charge of everyday meals." She goes on, "Only someone who does not need to make dinner *right now with a toddler attached to their leg* has the luxury of turning it into a weekend-tinkering pursuit."[96] A modified version of this critique appears to emerge with hobbyists using robots as part of participant fatherhood. As in the previous example, robots can help fathers practice nurturing and develop programs to benefit their children, but the time spent setting up these robots might displace and distract from the care they are supposed to assist with. Caring for a child using a robot would seem to be a privilege only accorded a parent whose time at home was not monopolized by meeting more basic needs. Still, robots differ from applications like recipe software that might be critiqued for trying to make domestic labor more masculine by turning it into a hacking project. In the case of programming robots, this technology also attempted the opposite, making programming labor more feminized and domesticated.

When programming and using robots, men's computer labor was recast as parental care—both to robots and to their own children. These cute and diminutive features of robots, which encouraged men to crouch and bend to interact with them, challenged the notion of technology extending men's dominion over domestic space. This represented a different model of computer interaction than the self-empowering approach more commonly imagined by computer hobbyists. Computer interaction was made more nurturing to fit better in feminized domestic spaces.

Personal robots offered a caring version of robot interaction rooted in domesticity, offering men tools to relate to their families through the computer. Designs, marketing, and discussions of robots for the home tried to imagine programming as care work. While gaining intimate knowledge of computing and robotics principles, users could program robots to carry out simple tasks related to family life and childcare, like waking children in the morning, announcing birthdays, and playing games. This vision of programming labor promised to assimilate men's interest in computing with expectations of participant fatherhood and masculine domesticity—both through the activities men could program for and with their family, and because robots encouraged a kind of affectionate care. After all, robots were small, cute, and vulnerable to injury, which required that their users be attentive to their weaknesses and constraints.

Still, there were serious limits to the way personal robots might function as companionate technology. Although they incorporated men into tasks tied to family, these robots did not necessarily encourage users to contribute to other daily household labor. Like Arlie Hochschild has noted in her study of the second shift, when men did contribute to domestic duties, they opted for the more exciting, less routine, and less onerous labor.[97] Programming a robot to dance with your child or tell jokes would be just this kind of domestic duty.

Furthermore, when considering the accounts of men's participation in fatherhood through their use of robots, a specific version of fatherhood and domesticity emerges. The figures that populate these articles are confined to a narrow band of professional careers. Reviews of personal robots describe software development professionals, science journalists, child development specialists, computer teachers, college professors, and tech consultants. It becomes evident that using robots in companionate family life was heavily dependent on having a job where working with computers was required. Although this was not at all a common economic reality in the 1980s, it does point to an emerging ideal of the flexible information worker that became more common in later decades.

Although the potential address of home robots may have been limited to a class of tech-adjacent professionals, these robots were still an intriguing effort to recast programming labor as something more akin to nurturing and childcare. Presenting it this way envisioned it as tied into the particularity of domestic space and offers a mode of robotics labor that might strengthen rather than replace or distract from companionate bonds.

3 "A DOLL THAT UNDERSTANDS YOU"

Computer Talking Dolls as Parenting Proxies

In anticipation of the 1987 holiday toy buying season, Worlds of Wonder, the toy company famous for the Teddy Ruxpin talking bear, aired a commercial for their new interactive doll Julie. This ad emphasized the doll's simulated awareness of its surroundings. In what looks like a preindustrial workshop resembling a setting from *Pinocchio*, an inventor puts finishing touches on the doll as it seems to come to life. Julie reacts to a bright light and appears to reply directly to the speech of those gathered around it. In fact, Julie was capable of limited speech recognition.[1] Distinguishing Julie from other talking dolls, the commercial describes its novelty: "Finally, a doll that understands you." The commercial concludes when the inventor, originally creating the doll for a competition put on by the king, sees how his granddaughter responds to Julie; he sacrifices his potential public glory to gift the doll to her instead.

Although the commercial presents Julie as if it were invented by a traditional craftsman and makes no direct reference to computing, Julie's remarkable abilities were made possible by advanced digital technologies. Worlds of Wonder was a Silicon Valley toy company founded by former Atari employees, and Julie was the result of their collaboration with the electronics giant Texas Instruments, responsible for engineering Julie's digital speech recognition. In this example, advanced consumer applications of computing technology were used to create a doll that seemed capable of recognizing and responding to its girl companion.

The juxtaposition of traditional toy and high-tech present in this doll may initially seem strange, but Worlds of Wonder was one of many toy companies in the 1980s that designed and sold computer-mediated dolls targeted specifically to young girls. Girls' relationships with these technologies were shaped by their embodiment in doll forms. They were

anthropomorphic toys that took on the appearance of young girls ranging from infants to preteens. Dolls have long been associated with girlhood play and functioned as tools to promote feminine socialization in a
variety of forms.[2] Like dolls from prior years, computer-mediated talking
dolls resembled babies to care for, friends to share secrets with, or slightly
older girls to emulate. Yet, unlike previous dolls, Julie and other computer
talking dolls sold in the 1980s featured combinations of digitized speech,
sound activation, and speech recognition made possible by microprocessors embedded in their doll bodies. They were also equipped with sensors
that reacted to touch, movement, and changes in the temperature and
brightness of their surroundings. Computer talking dolls were marketed as
toys that could really "understand" or listen to their playmates and respond
to what they heard.

Dolls were designed for girls but their promotion also targeted parents. Doll ads played on parents' perceived desires and anxieties relating
to childcare, which were made worse by contemporaneous discourses
about early childhood achievement and children's use of computers. The
increasing cultural prominence of working women also contributed to
parental guilt. In the 1980s, many professional middle-class parents, worried about their children's economic future mobility, bought goods and
services that might accelerate their children's intellectual and career development. These products were also meant to compensate for the perceived
social isolation of kids whose mothers were away at work.

In this context, computer talking dolls were offered as an aid to parents, especially mothers, by both educating and providing social companionship for girls. Although their high price tags meant these remedies
were available primarily to middle- and upper middle-class families, these
computer talking dolls served as attentive companions for girls. They were
also gifts parents could buy as proxies to help their daughters' upbringing and social development. Computer talking dolls promised mothers a
way to extend their parenting. These toys could assist with child-rearing
because they seemed to listen and respond to girls thanks to digital speech
and sound recognition technologies.

This chapter will discuss three prominent examples of computer talking dolls produced in the 1980s: Baby Talk (Galoob, 1986), Julie (Worlds of
Wonder, 1987), and Jill (Playmates, 1987). I argue that they were offered as
proxies for parenting by functioning as playmates that would model and
instill specific kinds of girlhood fit for the 1980s. These styles of young girlhood saw girls as capable subjects and envisioned how computer-powered

dolls could mix traditionally feminine forms of play and more apparently educational and active ones.

I refer to computer talking dolls as proxies rather than as replacements. They served as delegates of maternal socialization. In his work on media proxies, Dylan Mulvin defines the more general phenomenon of proxies thus: "Proxies are those people, places and things that we choose to stand in for the world, they play all sorts of different roles in making societies run and holding communities together. . . . The person, place or thing is a trusted delegate, standing in for something else, allowing us to displace the need for the real thing in order to accomplish some social, technical or political goal."[3] Mulvin focuses on media proxies like test images, which serve as stand-ins for the creation of standardized knowledge, but computer talking dolls fit a more general definition. Talking dolls functioned as proxies insofar as they were trusted delegates invested with the authority to represent mothers. As parental expectations and workloads increased, talking dolls could serve as authorized stand-ins to help socialize girls by providing companionship and education.

Even as these dolls were intended to relieve pressures on professional families, and especially career women, who were trying to meet rising expectations for childhood enrichment, the dolls' automated and mechanical qualities made some parents and experts uncomfortable. Computer talking dolls were offered as solutions to anxieties about childcare and as support for working mothers. However, some critics argued that these dolls' programmed nature could hinder children's spontaneous play and thus exacerbated worries about overly structured childhoods. Computer talking dolls threatened to expose anxieties stemming from the increasingly achievement-oriented child-rearing practices of the 1980s and the imperatives of an emerging neoliberal girlhood that affected even very young girls through the playthings their parents provided them. Critiques about girls playing with computer talking dolls, then, were also expressing concerns about certain models of child-rearing in which motherhood seemed to become more mechanical or results oriented.

Computer talking dolls show the influence of feminized culture on computing more obviously than many of the other examples in this book. Doll makers had young girls and their parents in mind as a primary audience for their products. Programmers and engineers that helped develop these dolls worked in conjunction with those in the toy industry who had a long history of marketing to girls. But the influence of women's and feminist culture in the production of these dolls also can be found in how they conceive of and relate to mothering and changing expectations for women's roles in and out of the home. Computer talking dolls were shaped by shifting norms for motherhood and expectations for childhood achievement.

As computing spread into homes in the 1980s, one of the common promises made to urge adoption was that early exposure to computers could help children's educations and make them feel at home with this new technology that they would later need for career success. Yet, the rise of computers, video games, and educational software fueled anxieties about a widening technological gender gap in which girls were left behind. This was exacerbated by the growing associations between video games and masculinity. Video games often relied on tropes and figures from boys' play cultures. In contrast, computer talking dolls drew on girls' culture and doll play traditions to offer an alternative form of computing experience more invested in domesticity and girlhood and integrated with social interactions in the companionate family.

Childcare and Parenting Anxiety: Raising Superbabies

In 1981, the computer magazine *InfoWorld* printed the results of their "What in the World Contest," in which the editors asked readers to share fresh and original ideas for the future of microcomputers. Mimicking the often-inflated rhetoric circulating at the time about computers and childhood education, the winning entry imagines a future in which the power of computers would even be harnessed for the betterment of infants. The author opens by suggesting that contemporary conveniences and human progress have resulted in the isolation of the modern baby, forced to spend too much time alone in the sterile spaces demanded by modern hygiene conventions. In response, she proposes a "computerized entertaining and teaching machine"—the Compusuk. The author describes a user for the Compusuk that is so young he would control the device by sucking a pacifier, which would trigger the machine to create stimulating sounds like white noise, heartbeats, and motherly voices. At six months, the baby could graduate to manual manipulation of the machine, at which time it would become a tool for education about topics like physics and astronomy, allowing the child "an active part in creating the material of his lessons." The author continues: "Another possibility of the Compusuk is providing automated mothering for children in day-care centers or institutions."[4] In this case, the device would not only be used for education but also to relieve the child's loneliness and provide maternal care by warming the child and rocking it gently.

After describing the potential educational and emotional benefits of the Compusuk, the author proceeds to speculate on the potential harms. Will babies given control of the device overstimulate themselves, necessitating parental controls? Will the device be programmed with an ideological slant that imparts biased educational values or omits whole areas of human development? And most alarmingly, would access to the Compusuk be equitable? The author fears that the price of such a device will lead to a widening wealth gap. Middle-class children will prosper because of the developmental benefits bestowed by the Compusuk, while the children of low-income families will be at an extreme disadvantage.

The absurd name of this speculative device might suggest that it was meant as a joke or a dismissive dig at the overhyped promises of educational technology. But the judges for the *InfoWorld* contest seemed to take it seriously. In the introduction explaining why this entry was selected as the winner, one of the judges posits that despite its unfortunate name—he proposes Electronic Crib as a better alternative—the device was a valuable potential application of computing. He compares the Compusuk to B. F. Skinner's air crib and speculates that parents would consider its merits and judge it fairly.[5]

In fact, as an application for computers that would educate even very young children, Compusuk is not all that atypical. In the 1980s, it was common to read about nursery schools exposing children of three or four years old to the Logo programming language or parents purchasing both real and toy computers for one- or two-year-olds.[6] Where the discussion of Compusuk was more unusual was its explicit musings on the controversial notion of "automatic mothering" duties. This device presented computers as proxies for maternal attention. Yet, although this association of the computer as substitute mother was atypical, here, too, the Compusuk was responding to contemporaneous anxieties about daycare, maternal presence, and childhood loneliness.

Such worries were typical reactions to changing conditions for parenting in middle-class families in the 1980s. As noted in prior chapters, women with children were entering the workforce at an increased rate. By 1986, 54 percent of women with children under six years old, the age group targeted by computer talking dolls or the Compusuk, were in the paid labor force, compared to 31 percent in 1970.[7] More families were being headed by single parents during this period as well.[8] These changes put a strain on juggling work and childcare for many families in which women were often expected to manage both. Additionally, with families averaging fewer children, there were fears that kids were experiencing greater loneliness and not getting the parental attention they needed.[9]

Concerns about lonely or neglected children were a widespread reaction to these shifts in women's labor. The figure of the latchkey kid, a specter that circulated previously in the 1940s, came to prominence again in the 1980s as a representation of anxieties about lonely, unsupervised children. Cultural commentators, childhood psychology experts, and public figures often raised the image of this burgeoning population of latchkey kids, named for the housekeys worn around a chain on their necks. As noted, this phenomenon was not new, and some experts even doubted the problem was actually getting worse. Nonetheless, it was often reported that in the 1980s six million children between ages five to thirteen were left alone to care for themselves after school, which was treated by many as a crisis in the making.[10]

This was exacerbated by inadequate daycare arrangements for many families throughout the decade. Whereas 21 percent of young children aged three and four were in daycare in 1970, that number reached nearly two-thirds by the late 1980s. Discussing cultural responses to the childcare crisis, Alice Leppert notes that despite more than one hundred daycare bills proposed in 1988 alone, substantial government support did not materialize.[11] Many parents were forced to rely on private daycare services for their childcare needs, and access was often unequally available to higher-income families who could afford it.[12]

Many parents, and especially mothers, struggled with anxiety that their time away at work would hinder their children's future well-being. Guilt about parenting contributed to a compensatory culture of over-programmed children, especially among the professional middle class. Experts suggested that working parents, and especially mothers, were putting pressure on their children to achieve as a reaction to their own feelings of inadequacy.[13] Of course, this can be understood as part of a backlash against feminist gains. But nonetheless, the pressure on women to be supermoms, confidently balancing career and childcare, was pervasive.[14]

Parents' anxieties about child-rearing stemmed from the urgent need they felt to prepare their children for a future economy in which there would be fewer opportunities. For professional parents, one emerging trend was "hothousing" children to ensure their success. Hothousing was a child development theory that posited that even very young children could be induced to learn and process advanced educational material through structured lessons and that this would encourage future intellectual achievement. A prominent proponent of hothousing, working out of the Better Baby Institute in Philadelphia, set up programs that taught parents to use flashcards to drill children as young as eight months old on math and reading.[15] Newspapers and magazines documented the rise of these "superbabies," describing children being shuttled from lessons in French, to gymnastics, to piano. Newspapers reported on two-year-old girls entertaining adults by naming the paintings of Vincent van Gogh; such publications also eagerly related stories about young violinists and other early achievers.[16]

Computers were implicated in the hothousing trend in multiple ways. Some experts suggested they could be used to educate even preliterate children, thus increasing those kids' potential head start. Computing was also viewed as one of the valuable skills for a changing economy that children could start to learn at a young age. Mothers could read in women's magazines and major newspapers about three-year-olds in special preschools learning Logo with the help of MIT professors or nursery schools using Apple II computers for graphics and music making.[17] In a special issue on preschool computing, *Family Computing* magazine featured a story about a two-professor family whose sons of two and four were using the computer for counting, drawing, and, in the case of the older child, word processing as well. If this were not enough to make parents worry that their children were not keeping up, the boys' father asserts that his children are not whizzes, but rather typical of preschoolers he encounters through his own research.[18] These stories contributed to concerns that children who weren't familiar with computers might be falling behind their peers.

Educational computing technology for young children was not limited to programming or software for microcomputers. Companies also created electronic toys and toy computers that could be used by children considered too young for a "real" computer. These devices also used microprocessor technologies, like microcomputers, but as dedicated toys they could experiment with different interfaces and voice synthesis for young children. In 1978, for example, Texas Instruments (TI) released the Speak & Spell, considered one of the first successful uses of digital speech synthesis in a commercial application.[19] Embodied in a handheld plastic case, the Speak & Spell used digital speech synthesis technology to create an educational toy to drill children on spelling. The Speak & Spell was modeled after "a cooperative friend or parent" that calls out spelling words.[20] Gene A. Frantz, one of the Speak & Spell engineers, would go on to help develop the TI digital signal processing technology that would be used for

Worlds of Wonder's interactive doll Julie. TI was involved in other computer toys throughout the 1980s, such as the Computer Fun or the My Little Computer.

Toy computers were also produced by Sears, Vtech, and Fisher Price. Mattel debuted their Teach and Learn Computer in winter 1982, promising parents "a smarter kid in 60 days, or your money back." Mattel advertised that the toy would serve computer-curious three- to eight-year-olds by addressing children "in friendly human-sounding voices."[21] A media relations manager for Texas Instruments explained that computers for preschoolers fostered classroom success and eased any potential tech anxiety, helping children "make friends with computers." A woman buying one of TI's toy computers for her two-year-old told a reporter she hoped it would help her daughter gain an advantage in her nursery school applications.[22]

Although popular with professional parents, critics contested the benefits of hothousing. Child development psychologists pointed out that hothousing had little scientific backing.[23] Others argued that middle-class parents were misapplying studies that had explored the benefits to economically disadvantaged children of being placed in early Head Start programs.[24] Middle-class parents used this research to justify spending money educating their own children at earlier and earlier ages to ensure their future advantage, with no evidence of benefit. Moreover, they were ignoring the interest in social equity that motivated Head Start programs in favor of a single-minded focus on individual achievement. This reflected what historian Victoria Cain argues was a larger shift in educational policy to increasingly define schooling as "equipping individuals for participation in a meritocratic global market" rather than seeing it as a public good.[25] Meanwhile, the unequal access of middle-class children to computers, both at home and in school, was threatening to further entrench inequality.[26]

Another critique of hothousing concerned the quality of learning and the values imparted to children subject to early education. Critics condemned the focus on memorization and rote learning as a corruption of childhood that would inhibit play and creativity. Psychologists pointed to the increase in stress-related disorders that children were exhibiting.[27] Experts and commentators announced that such pressures were leading to "hurried children" and the "disappearance of childhood."[28] Concerns about hurried children often pertained to fears about precocious delinquency, drug use, and sexuality for young teens; for younger children, these fears were related to anxieties about children's play. Critics pointed out the importance of seemingly unproductive play to healthy develop-ment and saw the bypassing of childhood as a harmful trend.[29] News-papers and magazines reported on superbaby burnout, as younger and younger children were being treated for headaches and stomach pains associated with anxiety.[30]

This backlash also implicated computers. There were fears that start-ing children on computing too early would lead to undue stress and an overemphasis on memorization.[31] Some experts were worried about edu-cational software's emphasis on drill and practice, arguing for more cre-ative uses of computers. Another concern was that computers lacked the tactile richness of other forms of play, such as with building blocks or digging in sandboxes. Others feared that computers could lead to social isolation at an age when it was important for children to learn to be around others.[32]

The question of whether computer education at too young an age would lead to robotic children energized critics. Discussing the emphasis on computer flashcards, computer commentator Fred D'Ignazio mused, "One wonders what a kid who gets computer flash cards at three months is going to be like when she gets to the ripe old age of five years, or ten, or fifteen. She may have a lot of computer facts under her belt, but how well-adjusted will she be?"[33] D'Ignazio was a proponent of computers for young children, but not when approached through drill and practice. Cri-tiquing programs at the Better Baby Institute, one psychologist argued that these attempts to force young children into formal learning "turns [them] into little computers."[34] Another critic of hothousing argued that it led to "overprogrammed children."[35] One mother, describing her amusement trying to use the flashcard method with her thirteen-month-old child, dis-cussed how this process made her daughter "a new computer in a world of typewriters."[36] The desires to assure children's achievement, structure their time, and force their play in productive directions risked producing mechanized children who learned only by rote.

Computers' educational potential raised hopes, but concerns about gender gaps, class disparities, and overprogrammed children loomed. Girls were in the classrooms using Logo and experimenting with programming, and they popped up occasionally in accounts of these efforts.[37] But the gender gap in computing was nonetheless seen as a major issue through-out the decade. Commentators described how girls were often less en-gaged with computers or how conditions in the classroom or at home discouraged their computer use. As Morgan G. Ames describes in her history of One Laptop per Child and its precursors, many of these computing education initiatives, such as Seymour Papert's work with children using Logo, were based on social imaginaries of what Ames calls "the technically precocious boy." Even though these experiments with educational computing were not explicitly gendered or designed to exclude girls, they were often inspired by the overwhelmingly male MIT hacker culture and based on ideals for childhood associated with boyhood. The imaginary child using computers was "characterized by seemingly innate creativity, fearlessness (whether with physical feats or technology), innocent mischief, and 'rugged individualism.'"[38] In the absence of explicit attempts to include girls in this vision of computing, Ames argues, this association with rebelliousness and competitiveness continued to marginalize many girls who are socialized with different values and expectations for their behavior.

There were some explicit efforts to include girls in computing in the 1980s, albeit targeting teens rather than the younger audience for talking dolls. The East Harlem school district in New York sponsored a daylong conference on "Daughters, Parents, and Technology," at which Seymour Papert was a keynote speaker. This event was meant to address the low numbers of teenage girls who continued beyond mandatory computing classes in school.[39] Computer camps for girls were offered to encourage those in junior high and high school, schools undertook studies to understand why girls were avoiding computer classes and offered strategies to counter the perceived masculine associations with computers, and organizations like the Girls Clubs of America sought grant funding to form computer clubs for girls.[40] In a rare instance of software made specifically for girls in the 1980s, developer Rhiannon created a few educational programs with girl protagonists such as *Jenny of the Prairie* (1983), suggested for users between the ages of seven and twelve. Rhiannon advertised their software as fun and educational, but also as a way to familiarize girls with computing.[41]

Computer talking dolls were part of this larger ecosystem of toys and services that catered to middle-class professional parents worried about their children's social isolation and future career achievement, particularly for girls. Toy companies had these anxieties in mind when they promoted dolls like Baby Talk, Julie, and Jill as complex, enriching toys for girls. Yet, computer talking dolls held an ambivalent place in the discourse of educational computing. They were touted for their high-tech abilities, but toy companies rarely labeled them as computer toys or claimed they fostered familiarity with computers.

There are some exceptions. Julie, for example, appeared on an episode of *The Computer Chronicles*, a PBS show devoted to computer software and hardware. The hosts admitted that the doll featured much of the new technology they had been discussing on the show all year: speech recognition, motion detection, and light and heat sensors. But computer talking dolls were not framed as ways to foster computer literacy or to familiarize girls with computing.

Admittedly, these dolls would not teach programming skills, but they were not much further from "real" computing than were toys like the Sears Talking Computron or Mattel Teach and Learn Computer. Yet, the difference does not appear to be entirely dependent on the gender of the audience for these toys. Preschool girls were commonly featured in advertisements for toy computers and in articles about parents desperate to buy computer toys for their young children. It is possible that computer talking dolls were less likely to be categorized as computers because of the voice interface they used to register the inputs of their girl playmates.

Other computer toys also talked, but unlike dolls, they required children to input responses through button presses. The same speech recognition technology that made talking dolls so advanced may have also made them seem less like computers of the period. The friendly and conversational mode of interaction allowed by computer talking dolls may not have squared with common expectations about what computer interaction or literacy would look like.

Still, these dolls were touted for their technologically enhanced abilities to talk and listen to young girls as proxies for parents. They represented an application of computing technology that catered toward conversational care and attention while also making gestures to educational enrichment.

Dolls That Listen

Microprocessor dolls were part of a wider toy industry trend of electronic toys since the early 1970s. Video games were the most prominent application of microprocessors to children's play during this period, but their fluctuating popularity affected other sectors of the toy industry. In the late 1970s and early 1980s, toy companies exploited the video game trend by investing in electronic toys. A number of handheld electronic games were released in the late 1970s—such as Parker Brothers' Merlin and Code Name: Sector; Mattel's Electronic Football, Auto Race, and Missile Attack; and Milton Bradley's Simon. These games and toys consisted of portable plastic packages embedded with dedicated microprocessors. Many of these games featured sound and LED outputs and press-button inputs. As Michael Z. Newman argues, the promotions for these toys situated them as computerized playmates and friendly competitors; they were thinking machines given agency by their digital components.[42]

The industry enthusiasm for electronic toys was tempered by the 1983 video game crash. But with the rebound of video games after 1985, other sectors of the toy industry reintroduced computing and electronics into their toys. For toy makers, this was further encouraged by attempts to replicate the surprise success of the animatronic talking Teddy Ruxpin toy in 1985 that sold almost $100 million in its first year.[43]

Whereas earlier dolls and bears, including Teddy Ruxpin, reproduced voice and music with the aid of miniature phonographs or cassette tapes, many new dolls in the latter half of the 1980s instead used digital technologies to make dolls speak and listen. A year prior to the release of the Julie doll described above, Galoob launched their Baby Talk doll. In 1987, Playmates released their talking doll, Jill, which combined microprocessors with mini cassettes. Beyond these computer-mediated talking dolls, there were a wide array of talking anthropomorphic toys produced during this period, including both baby and adolescent dolls as well as talking bears and other animals. Other instances included Axlon's AG Bear, Mattel's Baby Heather, Worlds of Wonder's Pamela, and Coleco's Talking Cabbage Patch dolls. This chapter focuses on Baby Talk, Julie, and Jill because they exemplify a range of interactions with computer-mediated anthropomorphic dolls targeted specifically to young girls.

Toymakers in the 1980s relied on many preexisting associations with doll play when developing computer talking dolls. Dolls have long been important in play as instruments for instilling normative feminine values and behavior in young girls. In her history of doll culture, Miriam Formanek-Brunell offers an account of changes in doll design over the course of the nineteenth and early twentieth centuries. Dolls transformed from tools for teaching practical domestic skills such as sewing, to models of female gentility that emphasized fashion and feminine display, to objects to instill more maternal and caring behaviors in later decades. Differing conceptions of ideal femininity and girlhood lead to varied strategies as adult culture seeks to define and utilize dolls for socialization in girls' play.[44]

At any moment, too, different types of dolls exist that encourage various types of play and styles of girlhood. Scholars have also argued for the complexity of this socialization process. Girls' relationships to dolls take on many forms. Dolls can serve as role models, figures to care for, or outlets for fantasies and desires, often deviating from parents' or doll makers' expectations.[45]

The way computer technologies were incorporated into talking dolls was shaped by this history. At the same time, doll creators sought to define unique capabilities made possible by microprocessors to justify the use of these complex technologies and the toys' higher prices. Given the complexity and expense of using computers to synthesize speech and recognize voice and other inputs, what advantages or special capabilities did microprocessors make possible and how might they allow toy companies to better appeal to parents and children? And for parents buying these toys for their daughters, what type of girlhood would computer talking dolls help develop? The dolls were designed with computer technology that allowed them to react to, and ideally channel, girls in their play—in advertising language, these dolls listen, know, and understand their playmates. With this capability, toy makers sold these dolls as more effective proxies for maternal care that could shape girls' play in directions deemed productive and enriching.

To some extent, dolls have always served as proxies in children's play. As tools for feminine socialization, dolls are invested with the authority to act in place of direct parental attention and help communicate culturally approved ideas about girlhood. Microprocessor technology enhanced toy makers' claims by promoting educational enrichment along with social companionship. Speech recognition and other sensors also added to the impression that dolls were aware of their playmates, allowing them to more effectively stand in as proxies for parental care. Busy professional mothers could feel that computer talking dolls would engage children in enriching social play experiences, teaching girls about feminized behaviors like maternal care or friendship while also modeling active and powerful interactions that appealed to parental desire for play to be educationally enriching and help girls gain a developmental advantage.

Baby Talk was most like a traditional baby doll. Its digital technology was mobilized to encourage maternal care from girls. Yet, this baby doll was also presented as a potential learning tool, and Galoob planned explicitly educational accessories to augment its ability to engage children. Julie was framed more as a friend and an equal that provided partnership, and the doll's computer technology allowed it to model cooperative turn taking. By playing with Julie in this way, girls could experience themselves as more capable as they saw their commands translated into play. Jill was closest to a fashion doll, old enough to guide girls through more active and precocious pretend play but still a largely innocent preteen. Its speech recognition gave girls the ability to control the direction of the doll's elaborate tales. The computer technology in these dolls was leveraged to foster more active and seemingly powerful play that aligned with parental expectations for educational enrichment while still representing norms of white girlhood femininity.

In all three dolls, their ability to simulate awareness with the help of microprocessor technology contributed to their work as maternal proxies because it automated the process of listening and reacting to girls in their play. Yet, the appeals to educational play made with computer talking dolls differed from those associated with children's software or even preschool toy computers. At times toy companies made claims for their dolls that we would conventionally think of as educational—Galoob planned for Baby Talk to teach skills in learning the alphabet, and Worlds of Wonder incorporated a sensor that would allow Jill to simulate reading with a child. More commonly, though, it was the dolls' abilities to simulate social companionship that was cited as their most unique capability for enriching play.

In many ways Galoob's Baby Talk was typical of the style of baby doll that had long been sold for girls' play, but it featured sound activation and other sensors to reinforce the caretaking behaviors associated with this category of toy. Baby Talk was one of the earliest examples of a commercial talking doll with speech provided by microchips. It could speak sixteen different phrases, accompanied by synchronized eye and mouth movements.[46] The doll resembled a blonde-haired, blue-eyed infant girl, although future releases of the doll included a black infant girl and a white infant boy. In 1986, Baby Talk debuted at sixty-nine dollars and became one of the best-selling toys that holiday season.[47]

Parents purchasing Baby Talk were reassured that its digital technology instilled appropriately feminine and caring forms of play, helping the dolls better function as proxies. As Galoob claimed, "A talking, interactive doll makes demands on the child, requiring that responsibility and caring be assumed by the child as a mother."[48] But Galoob also relied on contemporary ideas that play needed to be more enriching and educationally productive. In fact, Galoob included a pamphlet with the doll titled

Helping Your Child Learn with Baby Talk (Figure 10). This document, which Galoob explained was developed with the help of several child psychologists, frames doll play as an important educational activity: "Toys serve to open children's eyes to the world and facilitate sensory awareness, sound, sight, touch, dexterity, coordination, problem solving, memory, language development/understanding and important social and creative skills."[49] Baby Talk was presented as a toy that could train girls in caretaking while also developing other cognitive and developmental skills.

Figure 10. Front cover of Galoob's pamphlet *Helping Your Child Learn with Baby Talk* that was packaged with some editions of the doll.

With Baby Talk, Galoob utilized digital sound activation technology and other sensors to reinforce caretaking behavior by enabling dolls to respond to girls' input. By automating feedback in this way, digital technology augmented the doll's ability to encourage maternal play. Baby Talk's speech consisted of expressions of affection and requests for physical care including "Hug me," "I'm hungry," and "I like to be picked up." Baby Talk was also fitted with sensors that enabled it to offer targeted responses to being moved or fed a small bottle that was included with purchase. For example, when Baby Talk is on its stomach, it asks to be turned over and will repeat this request if the girl does not oblige. If Baby Talk says it is hungry and the child responds by placing a bottle in the doll's mouth, this activates a simulated sucking movement and sound from the doll. If the bottle is removed, Baby Talk asks for more, encouraging its "mommy" to continue feeding it.

In both these cases, the doll responds with targeted phrases of reinforcement or feedback when the child performs the expected caretaking behaviors. Baby Talk spoke similar phrases to previous talking dolls like the pull-string Chatty Cathy dolls popular in the 1960s and 1970s that also issued requests for care and expressed affection for their playmates. But unlike the randomly selected phrases of prior dolls, which required physical manipulation to initiate speech, Baby Talk's sensors enabled it to provide targeted positive and negative feedback for desired and undesired actions, more explicitly reinforcing caretaking behavior.[50]

For Baby Talk, the computer technology largely works to reinforce and automate long-standing maternal nurturing tropes for girlhood play. As a result, Galoob could market the doll as a more effective proxy that parents could give to their daughters to channel their play in desired directions. The pamphlet packaged with Baby Talk even included ideas for how the doll's interactivity could be optimized. According to Galoob, these activities "provide a level of structured learning to make the play experience more meaningful for everyone. These activities are designed specifically to enhance doll play, especially with the new interactive doll, Baby Talk."[51] Galoob offered specific exercises to structure the child's play meant to enhance capacities such as "nurturing skills," "sensitivity/feeling development," and "building sense of self." The activities suggested that parents might quiz their children on the appropriate caretaking response to Baby Talk's requests. Another activity, recommended for girls ages six to eight, was meant to help them analyze the mother–daughter relationship. The pamphlet explains that if Baby Talk asks to play peekaboo but the girl can't play at the moment, parents should ask the child what is the best way to respond (e.g., "Mommy's busy now; we'll play later").[52] Interestingly, in teaching girls nurturing responses to Baby Talk, the activity also encourages understanding their own mother's reactions to play requests.

In addition to encouraging caretaking and developmental skills, Baby Talk's function as a maternal proxy was extended by its sound activation, which urged girls to take on a more vocally active orientation to their doll. Getting Baby Talk to speak requires girls to talk—or at least to generate sharp noises that doll sensors would interpret as speech. As long as the girl keeps talking, Baby Talk will stay animated and continue to play with her. When the child stops speaking or making noises, Baby Talk begins a winding-down process, first expressing that it is "sleepy" and then turning itself off if no further speech is detected from the child. This mode of interaction not only mimicked conversational exchange but also encouraged girls to actively participate and express themselves to keep an exchange going. Galoob pointed to this more active potential, claiming that play with Baby Talk was "an opportunity to practice control over their environment."[53] The doll's interactive sound interface was a launchpad for encouraging more active play.

Galoob had ambitions that Baby Talk would be marketed in more explicitly educational ways with the help of a planned Interactive Video System, including a "smart box" accessory that could be attached to a VCR to play special video tapes. The tapes and smart box would work together to allow the doll to converse with onscreen characters. This technology, a result of Galoob's partnership with electronics inventor Ralph Baer, debuted in 1986 to accompany a talking animal toy called Smarty Bear. The Baby Talk version appeared in Galoob's 1987 catalog for toy retailers, but it seems this version never hit store shelves. The planned interactive videos would enable the doll to take on additional educational functions, such as helping children learn the alphabet. The advertisement sent to toy retailers explains the benefits of the interactive video system using the language of empowerment: "When Baby Talk learns her ABC's, YOU learn the ABC's! And both of you will be SO PROUD!"[54]

In the patent for the technology Baer even proposed a more advanced version with an additional microprocessor, allowing the child to respond to onscreen or doll prompts using push-button inputs. In a memo from January 1986, Baer lays out a philosophy and "ground rules" for the addition of this microprocessor technology. He suggests that there must be "real reasons" for the inclusion of this push-button technology, such as "instant feedback on whether [the girl's] answer was right or wrong" and "automatic scoring" of her answers. He also proposed the possibility that microprocessors could be used to enable the child and doll to play board games against each other, with the screen conducting their play.[55]

These proposals intended to leverage the microprocessor in the smart box to extend the impression of doll awareness, framing it within explicitly educational scenarios that establish a disciplinary relationship with the child. Galoob was contemplating how to account for parental expectations and design a doll that could better function as an automated proxy to impart educational lessons. The company even imagined that this baby doll could stand in as an instructor for girls.

To promote Baby Talk in the Minneapolis / St. Paul region, Galoob staged two public events for children in daycare programs. One brought a group of children from a local daycare center to the airport, where they boarded a plane populated by Baby Talk dolls. Another was a doll birthday party at a local park where five hundred children from nearby daycares were invited to celebrate Baby Talk's birthday. Along with these events, Galoob donated $5,000 to a local childcare lobbyist organization.[56] Such promotions targeted to daycares show that the company anticipated their dolls might be most valuable to parents who could offer them to daughters when unavailable during the workday.

But the company was careful not to explicitly claim that dolls could serve in place of parental attention. In fact, Galoob warned consumers not to think of the dolls this way: "Though it will help parents interact more forcefully with their children, Baby Talk is not a replacement for the time spent with them. It is, rather, a tool to nurture and enhance the parent/child relationship."[57] This disclaimer suggests that Galoob recognized that time-pressed mothers would seek a doll as a proxy caregiver to their daughters, but the company was also anticipating criticism of their dolls as maternal proxies.

Released a year after Baby Talk, the talking doll Julie offered a more advanced version of a listening doll with the inclusion of some limited speech recognition technology. Julie could distinguish between a few different utterances from the child and respond accordingly. It had the appearance of a young girl and was presented more as a peer or a playmate for the child, rather than a baby to take care of. As a maternal proxy, Julie offered to orchestrate a more active model of interaction in which the doll and child guided each other through different play activities. For example, Julie might ask to play pretend or have a party, and it came with books that described additional playful activities like going to the zoo or holding a fashion show. The doll's speech recognition technology came at a steep price—Julie retailed for over $100—but this technology allowed it to function as a more customized and autonomous proxy that directed and reacted to girls' play.

Julie's speech recognition chip was designed to highlight the doll's supposed awareness of its playmate and to allow the user to guide play. For Julie, Worlds of Wonder worked with Texas Instruments to produce a doll with custom speech technology. Julie's digital signal processing chip allowed it to recognize a limited number of preselected words that would initiate specific actions or play scripts. For example, one of these preselected words is "pretend." When a child asks to play pretend, Julie will begin talking about pretend play and use some of its existing phrases related to this topic, such as pretending to be a duck or asking the child to growl like a tiger. Girls could initiate particular branches of play and get Julie started talking about desired topics by using these secret words. As described by TI engineer Frantz, speech recognition technology provided the potential for a doll that could "listen" as well as speak.[58] As he describes, one of the primary design goals was to create "a variety of play modes that could be guided by the child."[59] Julie's speech recognition allowed the doll to better simulate awareness of the child's input and to modulate play in a way that appeared to recognize the child's responses and desires.

Julie's speech recognition had the potential to make girls feel empowered, giving them control over playful interactions with the doll. The impression that Julie was responding to the child's desires was further pronounced in how it reacted to pauses in the child's input and changing conditions in the play environment. After only a few seconds of silence from the child, Julie will ask, "Are you still having fun?" and "Do you want to keep doing this?" It is likely this was designed to prolong Julie's battery life by putting it to sleep as soon as its playmate loses interest or leaves the room. The result is that the doll displays a sensitivity to the child's interest in the development of play and conversation, which creates the impression that the doll is primarily concerned with responding to the needs and desires of its playmate. This mode of interaction also added to the doll's effective performance as a proxy providing oversight of the girl's play.

Julie's monitoring also appeared to attend to the child's physical well-being. Julie's sensors could register when it was dark or cold, and the doll would utter, for example, "Can you see okay? It's kinda dark." In situations that might feel unpleasant for the child, Julie's sensitivity to these changes could make it seem like the doll was looking out for the user. Even though

Julie was designed to look like a young girl that players could relate to as an equal, this aspect of the doll's technology allowed it to function more as a proxy that takes care of the child.

In its marketing, Worlds of Wonder emphasized Julie's ability to simulate awareness or understanding as a proxy caregiver. In advance of the holiday toy season, Julie was promoted with an animated television special, *Meet Julie* (dir. Walt Kubiak, 1987), that provided a fictional backstory for the doll. In this cartoon special, Julie was created by a single father as a companion for his only-child daughter. The father, a computer security specialist, presents Julie to his daughter to keep her company during an upcoming work trip to Paris. The daughter and her doll have solo adventures on this trip apart from her father's care, and the doll helps look after the girl as they navigate the city together unsupervised. Julie, the fictional doll character, performs abilities like the real Julie doll. For example, as the girl is being chased by jewel thieves and runs out in the rain at night, Julie remarks that it is dark and cold. Here the doll shows that it recognizes the child is in peril. In other parts of the special, Julie uses its powers—this time imagined ones—to help save the child from dangerous situations. Although it is an exaggerated and spectacular scenario, Worlds of Wonder seems to suggest that like the fictional father in *Meet Julie*, real parents who feared they were not providing their children with enough attention could purchase Julie to give to their daughters as a proxy, thereby assuaging their guilt and anxiety.

Not only did Worlds of Wonder present Julie as a doll that understood and monitored the user to show care, the company promoted the doll in ways that emphasized its potential to work as a proxy maternal figure that could convey educational lessons. Julie was sold by Worlds of Wonder in packaging that prominently claimed Julie to be "the world's most intelligent talking doll." Julie was also equipped with sensors on its fingertips that allowed it to interact with special books that Worlds of Wonder sold alongside the doll (Figure 11). When Julie's fingers were placed on the "magic spots" in these books, the doll would read aloud a word or name the image beneath its fingers. Julie's ability to read with the child could make the toy seem more enriching, even as its accompanying books were another opportunity to sell accessories for the doll. Even for children not yet able to read on their own, Julie could function as a proxy to help with reading or recall. This simple educational activity resembles the drill-and-practice flashcards of hothousing, with Julie reading with the child when parents are unable. Gesturing to the intelligence of this doll was another way Worlds of Wonder could appeal to middle-class parents anxious to give their children a competitive edge. In addition to heightening the doll's simulation of companionship, this reassured parents that doll play was a productive use of time and that Julie was a trusted proxy to which they could delegate child-rearing.

In 1987, the toy company Playmates released the talking doll Jill. Jill, thirty-three inches tall, wore a trendy sweater and leggings combination with bright pink legwarmers, looking like a slightly older girl. Like Julie, Jill had limited speech recognition capabilities. But unlike Julie, the doll's speech was not digitally synthesized but rather reproduced on mini cassettes. Julie combined microprocessor and audio cassette technology—its mini cassettes held recordings of branching scenarios; at branch points, the doll would ask the child to choose between two or three options. Enabled by its speech recognition technology, the doll's microprocessor then selects the appropriate track and Jill would continue its storytelling having taken the child's input into account.

Playmates designed Jill as a doll that might be treated more as an aspirational figure for girls, an idealized version of a preteen. In addition to

Figure 11. Worlds of Wonder's Julie doll. The doll is shown with its box and a special book containing "magic spots" the doll's sensors could recognize. Photograph by author.

the outfit that came with the doll, Playmates sold six expansion sets, such as Jill Goes to the Mall, Jill's Cheerleading Tryouts, and Jill's First Job (as a candy striper). These sets consisted of a cassette that would "program" Jill to engage in a themed interactive scenario as well as an associated outfit and accessories—such as pom poms for the cheer set, a clipboard for the candy striper job, or a charge card for the mall adventure. Jill functioned as a proxy for child-rearing differently than did Baby Talk or Julie. The doll still served to instantiate the type of active play that parents were expected to provide or desire for their daughters, but it did so in the form of an older sister to emulate.

Jill's microprocessor technology enabled the doll to spin elaborate, interactive scenarios that the child could participate in and guide at several branching points. This made the doll appear to be a highly capable and largely autonomous playmate. For example, when the tape Jill Goes to the Mall is placed in the doll's back, Jill, after some exposition, begins to speak as if the doll and child are at the mall together. Jill points out and describes the many sights around them in this imaginary scenario. The doll goes into careful detail describing the cat poster in the store that says "Hang in there" and the "funny" shirt that reads "I'm too cute to study." The audio track even includes sound effects to contribute to the pretend play scenario that the girl and doll shape together, such as the chatter of mall patrons and the sound of a coin dropping into a wishing fountain. Jill's tech-enabled storytelling enhances the transporting quality of pretend play and the doll functions as a partner in this activity.

Jill, a voluble playmate, represented an energetic, talkative version of girlhood femininity that could serve as a proxy for the influence of an older girl. Jill was graded for ages four and up. For girls around this preschool or early elementary school age, especially those without older siblings or playmates at home, Jill simulated aspirational and adventurous activities—like babysitting and slumber parties—guiding them on the proper way to behave in these situations. For example, Jill tells the child that lots of kids hang out at the mall—"boys, too"—and thus suggests that it is important to get dressed up before they go. Jill offered lessons in girlhood that would mimic those that girls might receive from an older sibling.

Jill was modeled as an active, even precocious girl, but the doll made less pretense to function as an educational tool in a conventional sense (e.g., teaching reading or math). In the TV commercials for Jill, in trying to explain its speech recognition, the doll announces to potential girl users that it is because "I'm smart, just like you." Yet, Jill's intelligence names its technological capacity, social awareness, and even physical abilities to move in synchronization with its voice, but was not directed toward explicitly educational subjects. Whereas Galoob spoke to parents through their educational pamphlet and Worlds of Wonder invented a fictional story in which Julie was created by a father for his daughter, parents did not figure as prominently in Playmates' promotional material for Jill. Still, a marketing VP for Playmates suggested that these dolls were for the type of parents who spend on luxury goods for themselves and would extend this indulgence to their children.[60] Jill's high price tag made it an exclusive gift to pass between parent and child. But once given to children, the doll acted more like a role model to the girl user rather than a caring maternal presence.

On the one hand, Baby Talk, Julie, and Jill were iterations of preexisting modes of talking doll play, only they used digitally synthesized or directed speech. On the other hand, these dolls' computer technologies were primarily devoted to increasing their listening and responsiveness rather than to speaking. Baby Talk and Julie did not have a much larger vocabulary than pull-string dolls popular in previous decades and spoke many fewer different phrases than cassette-based toys like Teddy Ruxpin or Cricket popular earlier in the 1980s. Jill was the most talkative, but this is because the doll combined tapes and microprocessor control.

What was most unique about these dolls was that they could listen and respond in more specific ways to their users and thus perform as more convincing proxies for parental presence. Of course, this was not unprecedented. In literature and fantastic stories, long before computers made it possible for devices to sense their environment, dolls had been depicted as toys that monitored the behavior of the girls who played with them. As Eugenia Gonzalez has shown in her survey of nineteenth-century doll literature, fictional dolls often gained the ability to speak, which they used to expose the misbehavior or unfeminine deeds of their playmates.[61] In doll literature, a doll's ability to talk was connected to its omnipresent surveillance of girlhood. The sound activation and recognition in Baby Talk, Julie, and Jill worked to literalize this sense that dolls could stand in for maternal presence and monitoring. However, their use was mobilized to different ends than the demure forms of play encouraged in doll literature. Rather than secretly listening to divulge a girl's secrets and opening her to punitive discipline from parents, these dolls seemed to listen and sense their environment to demonstrate their companionship and attention. They were disciplinary in the sense that they were incitements to certain kinds of active and enriching play rather than tools for prohibiting play deemed inappropriate.

Like with analog dolls, the digital doll as proxy facilitated socialization. Rather than a passively obedient femininity, these dolls promoted an active version of girlhood aligned with neoliberal subjectivity. Scholars have discussed the shifts in girlhood representations and ideals of subjectivity coinciding with the emergence of neoliberal economics and postfeminist politics. Angela McRobbie argues that girls have come to be situated as the privileged subjects of neoliberal economic changes, "endowed with economic capacity" and "charged with the requirement that they perform as economically active female citizens."[62] As McRobbie explains, "The production of girlhood now comprises a constant stream of incitements and enticements to engage in a range of specified practices which are understood to be both progressive but also consummately and reassuringly feminine."[63] Similarly, Anita Harris discusses the emergence of a figure she calls the "can-do girl"—a capable subject fit for the changes of late modernity and deindustrialization.[64] This scholarship often focuses on popular culture in the 1990s and tends to discuss girls older than the preschool demographic for computer talking dolls. Still, the description of girlhood subjects as active and capable while still fulfilling expectations for femininity applies as well to play with computer talking dolls.

Of course, for girls of four or five years old in the 1980s, the aesthetics of neoliberal girlhood were not identical to that of trends like the girl power feminism popularized in the 1990s and later. Nonetheless, a nascent postfeminism is apparent in these toys given to young girls by professionally striving parents. When playing with computer talking dolls, girls were addressed as subjects of capacity, children who were expected to work on developing their skills and engage in optimized doll play. This would ensure that they did not fall behind on their way to future educational and career success. Furthermore, computer talking dolls could help mothers ensure that their daughters had a better chance of maintaining middle-class membership in a flexible and changing information economy.

The neoliberal girlhood produced by talking dolls was also presumed to be a white girlhood. Granted, in addition to white, blonde versions, both Baby Talk and Jill were sold in versions as Black dolls as well. I have not found any indication that Worlds of Wonder produced a Black Julie doll. The Black versions of Baby Talk and Jill were modified to have different skin, eye, and hair color. Baby Talk was also given a different hair style— the baby sported a simulated version of a curly natural style that differed from the white Baby Talk's straight hair. Otherwise, the dolls were identical, including their voices.

Galoob and Playmates did little to conceal the fact that they took the white dolls as the standard version. When buying a Black Baby Talk, for example, the doll's accompanying instruction manual and other material featured the white doll. Galoob and Playmates at least featured Black dolls in images on the boxes in which they were sold, along with depictions of Black children playing with the dolls—in nearly identical poses to their white counterparts. Ann duCille has discussed how toy companies, when designing and marketing Black dolls, often reduce racial and cultural difference in dolls to modifications in skin color or fashion, thereby reproducing difference as sameness and presenting "dye-dipped versions of archetypal white American beauty."[65] As duCille insightfully unpacks in the case of Mattel, creating culturally specific Black Barbies poses the challenge of determining a successful version of racial or cultural specificity. Doll makers risk reducing difference to a few legible markers.[66] Nonetheless, despite the complexity of this task, companies like Galoob and Playmates showed little interest in engaging Black consumers with anything more than minimal efforts. For Baby Talk, Julie, and Jill, it was expectations for white girlhood femininity that determined the dolls' design, outfits, and narratives.

Programming Play

Dolls like Baby Talk, Julie, and Jill, by appearing to be really listening to children, could speak to the powerful anxieties related to child-rearing circulating in this period—both regarding parents' fears of children's growing loneliness and pressures on parents to maximize children's future success. Talking dolls embedded with computer chips were promoted as educational and active while still representing norms of white girlhood femininity, but these attempts to discipline play elicited an ambivalent reception. Dolls brought out tensions that arose from child-rearing efforts to shape young girls into feminine subjects of capacity. The possibility that dolls could serve as proxies for child-rearing was met with critique, creating unease about the child-rearing philosophies and style of parenting that made such toys seem necessary or desirable.

Reviews and commentary on talking and listening dolls were often critical of the intimations that dolls would be proxy caregivers and educators for young girls. These critiques were also condemnations of particular types of parenting or mothering. One journalist reviewing these dolls described what she saw as their "sad" appeal: "Many of these creatures are marketed with the generic label *educational*—a word that ought to be federally regulated like *lite*. I have the uncomfortable feeling that some parents really do think that they are giving their children someone—else—to talk to." She goes on to critique companies like Galoob that present their talking toys as playmates "in an era when children have more dolls than siblings or even friends."[67] Other critics skeptical of interactive talking toys referred to them as "electronic babysitters" and even "surrogate parents."[68] As these commentaries suggest, critics understood these dolls to be responding to the increasing pressure on parents, especially mothers, to provide educationally enriching play within a context in which children might also be experiencing less parental attention or social interaction. Parents were criticized for turning to dolls as proxies for their own attention and presence.

Critics of dolls like Baby Talk, Julie, and Jill were not only concerned that they were deficient stand-ins for parental attention. Some felt that computer talking dolls were emblematic of overly structured or programmatic play—a sign of what was wrong about a new style of parenting that saw every aspect of the child's schedule, including playtime, as subject to optimization. The application of digital technology in computer talking dolls worked to emphasize the impression that dolls could recognize the desires of their users; but for this to work girls were required to modulate their play practices and their speech to be legible to the digital technologies and sensors facilitating these interactions. The inflexibility and programmatic nature of this regulated form of play exacerbated critiques of both these dolls and the rigid child-rearing philosophies endorsing them. In a sense, critiques of dolls stood in for critiques of parenting.

Even as computers made it possible for a girl to animate and engage her doll companion with her voice, they also put constraints on the type of speech that could be recognized and thus guided girls' play in particular directions. The failure of these doll proxies to work flexibly also drew attention to aspects of 1980s child-rearing inspired by inflexible, achievement-oriented goals. Baby Talk was equipped with a sound-sensing circuit that was activated by a burst of speech followed by silence.[69] This interface requires the user to speak deliberately and pause frequently to encourage her doll to keep responding, resulting in a mechanical-sounding back-and-forth between child and doll rather than the spontaneous outpouring of speech associated with children's play.[70] Galoob seemed to anticipate that the mode of interaction encouraged by Baby Talk would worry parents and critics. In its instructional pamphlet to parents, Galoob warns, "Children of all ages are fascinated with toys. Don't be surprised if children initially spend a great deal of time just listening and responding to Baby Talk in utter amazement. When they become more familiar with Baby Talk, they'll begin to exert control."[71] With this warning, Galoob also speaks to concerns that dolls like Baby Talk would render children passive in front of the toy, similar to parents' worries about television consumption or the rote memorization associated with hothousing.

Julie's speech recognition technology also worked in way that might constrain girls' speech and make their play seem more programmatic. Each time the doll was reset or its batteries were replaced, Julie would require its playmate to go through a "programming" process so that the doll could recognize its "secret words" as spoken by that child. During programming, Julie engages in a script that asks the child to repeat such secret words as "yes," "no," "Julie," "hungry," and "pretend." When a user speaks to Julie, her speech initiates a targeted response only when it includes one of the secret words. The ability to guide play sessions with the doll came at the expense of the variety and spontaneity of speech. Although it is true that this speech-recognition technology likely did not eliminate all divergent speech, a girl using this computer talking doll is encouraged to constrain her speech to preprogrammed scripts for effective recognition and response.

Jill's design appeared most likely to draw critiques for limiting children. When playing with Jill, the child's speech was responsible for branching choices in the doll's stories. In this case, the child could have direct influence on the branching narrative that Jill presents by answering its questions, but her freedom to respond in varying ways is highly constrained to only two or three choices of words. This is different even from playing with Julie because Jill is so garrulous and continues to speak even if the child is silent or provides irrelevant responses to its prompts.[72] To play with Jill would require a child to sit in thrall to its breathless pace of storytelling to keep up. Even though the "choose your own adventure" style of Jill's storytelling would seem to give girls power over the play experience, Jill's nonstop talk had the strong potential to overpower users.

Critical reactions to computer talking dolls were a microcosm of the same anxieties being debated in response to superbabies and hurried children. Critiques of these dolls allowed for an expression of condemnation over an overly goal-oriented and automated form of mothering.

The digital technology in computer talking dolls could provide educational enrichment and empower girls' voices in play. But to realize this potential, girls needed to constrain their play and accept a disciplinary relationship with their dolls. This tension, arising from parental attempts to imagine girls as subjects of capacity, revealed contradictions and ambivalences, particularly in the drive to optimize and structure their play through hothousing and other educational strategies.

Similar to the way hothousing was criticized as a corruption of spontaneous play, a common critique of computer talking dolls was that they undermined girls' ability to engage in imaginative or active play. For example, one critic asserts, "Toys like this make you wonder if people trust childhood anymore. Talking toys serve to try to get kids to have a more limited dialogue with themselves."[73] Another critic refers to these as "spectator toys" and warns that they "squelch the child's imagination and limit play," encouraging children to sit passively while their dolls act.[74] Another article about interactive toys, including Baby Talk and Julie, invited experts both for and against to describe the dolls' potential to the readers of *Working Mother*. The magazine quoted a psychologist who consulted with Galoob on the production of Baby Talk arguing that the doll helps children with vocabulary, language, and problem-solving. The same story also features experts who suggested that, to the contrary, these interactive toys made children passive and led to an automatic or Pavlovian interaction.[75] Another commenter explicitly critiques these types of interactive toys as a sign that parents did not value play and always wanted their child's time to be structured.[76] Similar to debates about hothousing, discussions of talking dolls balanced the desire to give children the most advanced toys and the best opportunities against concerns that childhood was being corrupted by the pressure for children to be productive at all times.

Computer talking dolls helped draw attention to the contradictions in how children's culture was being shaped in an age of hothousing and superbabies. The programmatic quality of these toys threatened to expose the way achievement-oriented mothering practices were focused on turning children into a performance of intelligence through memorization and scripting. This is not to argue that these dolls were harmful or that there are preexisting natural and authentic forms of play that digital technology corrupts. Although play is often depicted as spontaneous, natural, and unregulated, it has long been the target of parents, child-rearing experts, and commercial media producers seeking to control and shape children's development. Media historians have documented the intersection of play and the regulation of childhood across various media, including cinema, television, toys, phonography, and even daydreaming practices.[77] In all these cases, adult culture attempts to promote active and self-disciplined forms of play while still relying on the assumption of an inherent relationship between childhood, spontaneity, imagination, and natural innocence—assurances of what Jacqueline Rose calls "the myth of childhood innocence."[78] In the critiques of computer talking dolls, this same discourse of inauthentic play is used to scrutinize mothers who are judged to have allowed achievement-oriented goals to affect their child-rearing practices.

Creating a Conversational Environment

The intersection of dolls and computing in toys like Baby Talk, Julie, and Jill arose from many simultaneous developments in children's culture. The toy industry in the 1980s was undergoing changes that contributed to an increasing gender polarization of toys and children's television programming. This period saw a rapid expansion of licensed characters and tie-ins between toys and television. Critics associated these toy–TV crossovers, also derisively called "program-length commercials," with the entrenched gender differences in children's consumer culture. These toy-based programs were separated into those that represented hyperfeminine fantasy worlds like *Strawberry Shortcake, Rainbow Brite,* and *My Little Pony* and hypermasculine action-adventure programs like *ThunderCats* and *He-Man and the Masters of the Universe.*[79] Anxieties about these gender-segregated toys were made worse by the popularity of video games in the 1980s, which were primarily marketed to boys. Like with computers, parents feared that girls would be hurt by their exclusion from video gaming, worrying it would hinder their development of spatial reasoning and logic skills useful in future education and careers.[80]

The toy industry's polarized gender address to children was part of an intensification of niche marketing and served toy makers' interests, but parents and critics were not wholly comfortable with these trends. In her analysis of toy-based videos for girls, like *My Little Pony,* Ellen Seiter interrogates the strong reaction from adults against these programs' hyperfemininity. As she notes, a by-product of the rise of these licensed toys was that young girls were at last being targeted on a large scale as a unique audience for television programming. Despite the negative critical reaction to these programs and toys, they did not require girls to identify with boy characters and masculine adventure stories, but instead spoke to anxieties and desires specific to girlhood. Seiter's response is to take seriously and read closely some of the most derided girls' programming. She explores the appeals being made in these shows and analyzes the way they negotiate complex emotional and social anxieties related to girlhood, as well as how they adopted elements of women's film and television genres.[81]

Following Seiter's example, this chapter has aimed to take computer talking dolls seriously even though they may seem like elements of a degraded girls' culture or, like toy-based videos, bald attempts to exploit trends in the toy and electronics industries. In fact, computer talking dolls represent a different model of relating to computers and a way of using computers as a support to parent–child relationships that differed from educational software or video games. This was a model of computing as proxy for maternal child-rearing more connected to the social life of the home. Significantly, it is also a version of computer play that centers the traditions of girls' domestic play rather than boyhood culture.

Thinking of computer talking dolls as proxies for maternal presence offers a significant counterpoint to the tendency to think of video games, another tradition of computer play, as a repudiation of domesticity and motherhood. Returning to the design and patent documents for Baby Talk and the interactive video system planned for the doll, we can see how this mode of computer interaction was conceptualized consciously as an alternative to video games. Here, it is important to highlight that Baer, the inventor that designed this system with Galoob, was better known for developing the pioneering Magnavox Odyssey in 1972, often described as the first home video game console. In his notes on the interactive video system that would have accompanied Baby Talk, Baer contrasts it with other technologies like video game consoles or the ROB robot peripheral for the Nintendo Entertainment System. He classifies the robot and video game consoles in the category of video games or video-assisted games but argues that the technology proposed by the Interactive Video System aims "to achieve an entirely new objective."[82]

Unlike video games, which Baer claims emphasize the player's control or manipulation of action occurring on screen, he argues that what makes the Interactive Video System unique is its focus on the doll or bear in the living room and its relationship to a conversational environment that spans the screen space and the living room space.[83] He proceeds to suggest a setup in which a girl, holding a talking doll on her lap, is linked into a complex domestic assemblage. Even when she is not speaking, the girl participates through her doll and is put into conversation with an entire room of animated appliances like the TV and VCR. In the patent, Baer even suggests that multiple dolls could be wired together, thus recreating an entire social set for the girl to engage with. As Baer describes this idea, "the result is a conversational environment with interaction taking place between the screen character or characters and the two animated [figures], as well as the participating child."[84] Although never realized, this amplified version of the imagined social interactions for computer talking dolls demonstrates that toy makers were conceptualizing an alternative mode of interaction from that fostered by video games—one that was continuous with the domestic.

Video games were often marketed in opposition to the feminized connotations of domesticity and to maternal presence. As Bernadette Flynn discusses, it was common for video game advertisements to promote the technology as an exciting escape from the home into worlds of speed and danger. In many commercials, the home literally exploded from the energy of the video game.[85] Michael Z. Newman argues that even as video games were domesticated in the late 1970s and 1980s as a medium that the family could potentially play together, they still drew primarily from sports, military, and adventure themes associated with masculine culture.[86] Similarly, in his discussion of video games as virtual remediations of backyard boy culture, Henry Jenkins argues that games are framed as an escape from or masculine transcendence of the confinement or perceived feminization of domestic space and a rejection of maternal influence.[87]

Dolls like Baby Talk, Julie, and Jill are examples of girls' computing that precede the self-conscious efforts of scholars and game designers to create a girls' computer game movement in the 1990s. As represented in the essay collection *From Barbie to Mortal Kombat* and in games produced by companies like Purple Moon and Her Interactive, the girls' game movement sought to create computer games targeted specifically to girls. Scholars and game designers questioned whether the girls' game movement was best served by more positive representations of women and girls in games; games with themes, settings, and genres stereotypically associated with girlhood; or social contexts that encouraged girls' access to computing. One of this movement's central questions was whether designers should affirm conceptions of girlhood taste that are often belittled or transform gender roles through computer game design.[88]

Toys like Baby Talk, Julie, and Jill have not figured in scholarly discussions about gender and computing, but they offer additional insight into the question of what girls' computing might look like. Talking dolls leverage the computer's power to simulate social awareness with playmates, making the technology useful by promising that it will serve as a maternal proxy. The examples in this chapter show that the digital technologies used in these dolls were not mobilized to enhance control over kinetic physical challenges but to create an illusion of a social experience, allowing girls to talk to their television sets or read stories with a doll playmate. Whereas video games were imagined largely in relation to histories of boy culture, computer talking dolls represent an example of computer play that instead draws on traditions of girlhood doll play, which have long been associated with domesticity. In these toys, computers are made to fit into existing traditions of doll play, reshaping girlhood activities and redefining what it means to relate to computers or engage in computing.

4 SEX AND THE SINGLES GAME

Adult Games, Cringe, and Critiques of Masculine Seduction

IN 1987, AN ADULT-ORIENTED GAME called *Romantic Encounters at the Dome* was produced for Commodore Amiga and IBM computers. The game publisher, Microillusions, promoted *Romantic Encounters* as "the ultimate text experience for 'sensual singles' (or those pretending to be)." In this text-based interactive fiction game, players enter a posh Los Angeles nightclub called the Dome, where they could meet many attractive singles (non-player characters) with whom to explore "a romantic rendezvous or erotic encounter."[1] Some of the women that players encounter include Tanya, a romantic looking to escape a domineering partner; Priscilla, just out of a relationship and interested only in a casual fling; Cathy, desperate for a meaningful connection; and Jeri, a thoughtful photographer who shares many of the player character's interests. To proceed through the game, players must determine what text inputs are required to keep each non-player character's attention. Not every character responds positively to the same approach. Priscilla is only interested in anonymous sex and will end the encounter if the player tries to engage her in conversation, whereas Jeri will kick the player out of her apartment if he tries to sleep with her too soon. While playing *Romantic Encounters*, users also learn about their own character's views and attitudes. Despite initially appearing to be a relatively desirable man, the player character is revealed to be quick to perceive slights, insecure about his masculinity, and overly instrumental in his relations with women.

When promoting *Romantic Encounters*, Microillusions claimed that it was more than a game: "Your experiences at the keyboard with *Romantic Encounters* can be carried over, if you dare, to broaden and enrich the rest of your daily life."[2] In addition to representing and simulating singles life, the game might help players reflect on their own views and habits in relation to seduction and sexual intimacy. According to Microillusions, then, *Romantic Encounters* was not a substitute or mere representation of singles life, but potentially a way to approach it more critically or with greater insight.

As computers were introduced into the home and became more widely available over the course of the 1980s, users and commentators were grappling with the role they would play in relation to larger renegotiations of masculinity and heterosexual relationships. Newspapers and magazines commenting on computer enthusiasm or even computer addiction focused largely on men and hypothesized that computers offered the pleasure of mastery and control to men who had trouble with the demands of interpersonal relationships.[3] The masculinity based on dominance and mastery attributed to some computer users was potentially at odds with ideals for masculine sensitivity and awareness circulating in popular discourses. Representations of sensitive New Men were appearing in newspaper and magazine profiles and editorials as a potential evolution of masculine norms—as were critiques of this version of masculinity in the form of headlines decrying the rise of wimps and wormboys, men who were perceived to have taken sensitivity to an extreme. At the same time, some feminist critics, especially antipornography feminists, drew attention to masculinity based on domination as a threat to women's well-being.[4]

Computers had an ambivalent relationship to these negotiations of masculinity. Was the excitement and pleasure of using computers a substitute for dating that catered to overly passive or overly controlling men looking to avoid the challenges of real romantic relationships? Would the feelings of mastery and control facilitated by computers contribute to a masculinity based on domination? And regarding the impact of these different styles of masculinity on women and other intimates in the orbit of these men, were computers an extension of a media culture that provided interactive control over the images of women's bodies? Or was it possible that relationships to computing could somehow translate to greater sensitivity and awareness about romance and sexuality beyond the computer, as *Romantic Encounters* claimed it would do?

Considering these contestations of masculine ideals occurring at the time, adult games like *Romantic Encounters* were a potential point of controversy for the computer industry as computer hardware and software were being adopted in middle-class family homes. Adult games, which handled mature and often sexual content, were released sporadically throughout the 1980s but evoked renewed attention later in the decade when a few such titles were released by more established software companies. In making and circulating these games, software makers invoked comparisons to pornography, adult media industries, and antipornography feminist critique, often to distance their own products from more graphic adult media. The resulting representations of masculinity and heterosexuality in adult games produced by these software companies were shaped by this engagement with feminist critique and with ongoing redefinitions of heteromasculinity. Rather than endorsing a sexually capable pornographic masculinity, these games represented versions of masculinity that were targets of critique and even ridicule. By criticizing these sexually promiscuous versions of masculinity, adult games were brought closer in alignment with companionate ideals and made suitable for the middle-class family homes in which they were likely to be played.

Focusing especially on *Romantic Encounters* and *Leisure Suit Larry in the Land of the Lounge Lizards* (Sierra On-Line, 1987), this chapter analyzes how adult games represented contemporary anxieties about heterosexual relationality and commented on shifting heterosexual masculine ideals. Contrary to fears that men turned to computers as an uncomplicated substitute for complex sexual and romantic relationships, these adult games represented sexuality as fraught with anxiety—and through these representations they interrogated and critiqued certain versions of masculinity while endorsing others.

The version of masculinity critiqued in each game differed. *Romantic Encounters* featured a player character that resembled the sensitive New Man of the 1980s, whereas the protagonist of *Leisure Suit Larry* was a failed version of a type of liberated playboy of the 1970s. Yet, the critiques of masculinity in both these games were managed through the utilization of cringe aesthetics. Julia Havas and Maria Sulimma theorize cringe as a conflicted mode—one that inspires moments of embarrassed recognition while also distancing viewers from the social transgressions represented.[5] The use of cringe in *Romantic Encounters* and *Leisure Suit Larry* encourages an ambivalent relationship to the player characters in these games and facilitates a position of reflection and critique for the presumed male user. Players were encouraged not simply to identify with the instrumental approach to picking up women the games depicted. Instead, the games were structured to put players in a conflicted position, causing them to reflect on the version of masculinity and relationality depicted. Although not explicitly presented as guides for how to, or how not to, pick up women, the games used computers to offer a critique of masculine ideals, even if this was largely done in a way that endorsed a different version of more monogamous heteromasculinity.

Video game scholars have offered some insights on the relationship between games, hegemonic masculinity, and a presumed heterosexuality. Discussions of gaming masculinity tend to focus on the way that avatars represent positions of power, either through physical prowess or control over other characters. It is common to argue that video game avatars and player characters function as masculine ego ideals through which players exercise mastery, but some scholars have analyzed instances in which games present more critical or denaturalized representations of masculinity, disrupting experiences of agency or control.[6] Still, these discussions rarely focus on the strategies through which games represent different styles of heterosexual relationality, and when they do, they often present them as conquest narratives.[7] The analysis of adult games like *Romantic Encounters* and *Leisure Suit Larry* in this chapter contributes to this scholarship by demonstrating how cringe aesthetics and affects offer another way to critique treatments of heterosexuality in games. Cringe works not primarily by endorsing or challenging physical control or avatar rationality, but rather by inspiring critique of conventions of social relationships—specifically, the way player characters relate to women.

Adult games may seem like a strange place to end a history of computing as companionate technology. Compared to the examples from previous chapters, adult games are less obviously connected to the computer's definition as a medium of companionate relationality in the 1980s. Although these games featured representations of heterosexual relationships, they appear to offer highly individual, erotic experiences with a computer. These were single-player games, targeted primarily to heterosexual men. They did not require a romantic partner to play like some of the romance software discussed previously. These adult games would not help with child-rearing or bring users closer to their wives. Unlike robots that cost hundreds or thousands of dollars and roamed around the house, these were inexpensive programs and likely not the type of purchase that would need to be justified as integral to family use. (Although, as I will discuss, it was still necessary to make a case that these were not harmful to the family.) In fact, there is even evidence that some men played *Leisure Suit Larry* on office computers, sometimes using illegitimate copies shared through informal networks.[8]

Still, in developing these adult games, software companies were conscious of the specific familial and domestic contexts in which they might be played. In circulating these games, companies considered the potential reaction of women and other critics as they worked to define the relationship between home computing and sexual and adult material. At the least, software makers found it necessary to make a case to consumers that these games would not be harmful to families or work as a corrupting force. Furthermore, even when played in isolation, these games could still critique masculinity and norms of heterosexual relationality, thereby affecting users' notions of relationships. Analyzing the creation, design, and circulation of adult games helps demonstrate how computers could function as technologies of domestic relationality not only through mediating individual relationships, but in shaping conceptions of relational ideals in companionate arrangements.

Nice Guys, Nerds, and New Men: Reshaping Masculine Ideals

In 1984, a book was published with the strange title *How to Make Love to a Computer.* Purportedly written by a computer author, Maurice K. Byte, this guide instructs men how to have a more satisfying sexual relationship with their own computers at home—or sometimes even that of an open-minded neighbor. The book makes many allusions to popular culture texts about post–sexual revolution culture and mimics the style and advice of experts and commentators that guided people through changing relational norms in the 1970s and 1980s. In a gag that runs throughout the book, for example, the author makes asides that refer readers to the many related (and fictional) self-help books that refer punningly to real bestsellers. *The Joy of Sex* (Alex Comfort, 1972) becomes *The Joy of Programming, The Hite Report* (Shere Hite, 1976) becomes *The Byte Report, Nice Girls Do!* (Irene Kassorla, 1980) becomes *Nice Computers Do!, The Sensuous Woman* (J, 1969) becomes *The Sensuous Computer* by "K."

Like the sex and relationship manuals it parodies, *How to Make Love to a Computer* unpacks myths and fears about men's sexual performance anxiety when interacting with computers, including size (of his hands) and stamina (at the keyboard); diagrams the erogenous zones of a "sensuous computer"; offers advice about the importance of romance through a case study example of one man's courtship with his desktop; and builds to a section on "The Joy of Programming," complete with illustrations of sex positions between a man and his machine lover. Throughout the guidebook, while discussing relations to computers the author addresses men as if they are disoriented by the changing gender norms in hetero-sexual relationships. Like the New Woman of the 1980s, the reader is told that the "new" computer has raised expectations of men. The new com-puter wants to be treated like an equal, and it desires both romance and sexual fulfillment from a partner: "Today's computer is learning to say NO without feeling guilty; it is tired of interfacing with partners who expect it to be a business computer in the office, a video game in the den, and a personal computer in the bedroom. This 'new' computer has much to give, but it needs to receive as well. Like you and me, it is an emotionally vulnerable partner that must be periodically pampered and reassured—a partner that is tired of being used, and ready to be loved."[9]

In *How to Make Love to a Computer,* discussions of men's relationships to computing were used to articulate anxieties about their relationships to women and express confusion about norms of contemporary hetero-sexuality in a post–sexual revolution and post–women's liberation con-text. The masculine figures that appear throughout this guidebook, as well as the presumed reader addressed by the text, are reflections of chang-ing ideas about masculinity. They are expected to be more sensitive to the emotional and sexual needs of their partners, informed about sex-ual techniques and open to experimentation, and aware of their own anxieties and hang-ups. But these figures are also presented in a joking manner. The men in this guidebook are seducers, but they are also huge nerds. The book has a tone of highly conscious self-deprecation. It knows that there may be something juvenile, pathetic, or even sleazy about the intense attachment to computers and the performances of masculinity it describes.

How to Make Love to a Computer is a particularly silly attempt to think through the relationship between men and computers and the impact the technology has on men's relationships with women—one that at times relies on misogynistic clichés about women as demanding and unreason-able. Yet, the book's conception of computers as substitute sexual part-ners and its comparison of computing to heterosexual relationships was common in the computing context in which it was published. In a more sober example, Sherry Turkle's ethnographic study of computer users from the same year, *The Second Self,* describes the relationship to computing in a similar way. Turkle suggests that hackers, as well as the new wider market of consumers adopting computers in the 1980s, were drawn to the devices as an alternative to confusing and complicated interpersonal relationships. She posits that users enter intimate relations with their computers out of a desire for more manageable pleasures and a preoccupation with control and mastery.[10] Along with many articles in newspapers and magazines reflecting on increasing interest in computing, these discussions were grappling with the intense feelings that computers could inspire in their owners and users, but were also considering how these attachments affected heterosexual relationships.[11]

As *How to Make Love to a Computer*'s many references to sex and relationship manuals suggest, anxieties about men's relationships to computing occurred within a wider context of changing ideas about masculinity and heterosexuality. In a 1984 *New York Times* article Barbara Ehrenreich described the prototypical New Man of the 1980s: aged between twenty-five and forty, affluent, single, and living in a city, the New Man is a style-conscious consumer of food, clothes, and home furnishings; obsessed with health and physical fitness; and, most distinctively, concerned that people find him "genuine, open, and sensitive."[12] Whereas successful masculinity for white middle-class men had long been firmly associated with marriage and breadwinning status, in the 1970s and 1980s men who avoided marriage and were dedicated to the pursuit of their own pleasure were identified as a new ideal.[13] The urgency for the New Man in the 1980s to wear the right clothes, eat the right way, and show the right amount of sensitivity and awareness was, according to Ehrenreich, a compensatory reaction to economic pressures threatening downward mobility. These precisely cultivated lifestyle choices were meant as signifiers of the New Man's belonging in the shrinking middle class.

In her discussion of the New Man as an affluent minority and a contested ideal, Ehrenreich critiqued the new masculinity not only for its consumerism and class associations, but also for its relationship to feminism and women. According to Ehrenreich, the new masculinity was not the result of men contending with feminist struggle or with their own privilege, but rather a "flight from commitment." Repudiating the expectations that they support a family, New Men instead pursued individualistic consumer pleasures—a less burdensome way of pursuing masculinity.[14]

Consistent with Ehrenreich's critique, other popular commentary equated the New Man's sensitivity with passivity and fear of commitment rather than a genuine transformation in gender relations. The columnist Deborah Laake referred to these men as wormboys. In an article in the *Washington City Paper*, reprinted in outlets around the country, Laake described wormboys as men exhibiting excessive passivity and fear of commitment and who lack ambition or political conviction, with weakly held opinions and beliefs and a susceptibility to feel threatened by women. According to Laake, these wormboys, or wimps, appeared at first as sensitive and emotionally intelligent, but were unwilling or unable to endure challenges when they arose, in relationships or in life.[15]

Other critiques of the new masculinity were not concerned with its implications for women, but instead reacted to a perceived feminization of men. Sociologist Michael Kimmel describes this period as "a virtual Great American Wimp Hunt," in which accusations of wimpiness were cast at film stars, public figures, and even presidential candidates.[16] The 1982 bestseller *Real Men Don't Eat Quiche* exemplified this anxiety—although it did so with a highly satirical tone. In this book, Bruce Feirstein lists the characteristics of "real men," whom he contrasts with wimps and quiche eaters, the "Alan Alda types—who cook and clean and *relate* to their wives."[17] Although the majority of Feirstein's attention is focused on critiques of the new masculinity's investment in consumer culture, this quote also indicates that shifting relations to women—in which men were less domineering and treated women as equals—was part of this identity.

But not all observers agreed that passivity was the largest problem plaguing the new masculinity in the 1980s. The flipside of this concern about passivity was a style of masculinity overly invested in control and domination of women. A result of the sexual revolution and women's movements of previous decades was a greater public awareness of the extensive role of male sexual violence in American culture. Feminist activists drew attention to the overly aggressive and violent tendencies of masculinity in patriarchal culture and the cultural prevalence of rape, sexual harassment, domestic abuse, and other indicators of male violence toward women.[18] This coincided with a greater mainstream visibility of pornographic media in the 1980s as the adult home video industry boomed. A subset of antipornography feminists emphasized the perceived role of pornography in fostering a style of violent male sexuality that depicted women as objects under male control.[19] For antipornography feminists, the threat of contemporary masculinity was not that it was too passive, but rather that it was overly invested in domination and control of women's bodies—as exemplified and exacerbated by pornographic media.

Where did computers and the budding class of enthusiasts fall into these contestations around changing masculine norms? The computer enthusiast, sometimes referred to as a hacker or nerd, offered a useful figure through which norms of a new type of white middle-class masculinity could be worked out, especially one based on intimacy with digital technology that was becoming increasingly widespread. These men's relationships to computing became a point around which to work through questions of passivity, control, and masculinity.

As scholars of masculinity like R. W. Connell have pointed out, men's use of computers, especially as part of professional careers, put into flux associations between man–machine relationships and physical prowess. The sedentary labor of computing and sitting in front of a keyboard shared associations with feminized labor. However, this feminization was combatted with the help of personal computer cultures that associated the technology with competition and power and even offered a cybernetically amplified mastery.[20]

Scholars discussing the history of representations of nerds have provided additional insight into this process of redefining masculinity in relation to computer technology. According to Lori Kendall, the figure of the nerd helped define a type of masculinity associated with technology that was previously liminal but came to be incorporated into hegemonic norms. Although that version of masculinity did not represent the masculine body as strong and powerful, it attempted to maintain domination over women, people of color, and working-class men and masculinities through cultivation of technological mastery.[21] For example, Kendall describes how films like *Revenge of the Nerds* (dir. Jeff Kanew, 1984) depict nerds as an oppressed minority in ways that mimic and displace discourses about Black civil rights. Ron Eglash discusses how representations of the nerd work to gatekeep technological fields so that they are made accessible primarily to white heterosexual men—although he claims it is not always successful in that project.[22]

In many ways, 1980s computer games aligned with this effort to code computer use as a white masculine pursuit and to reassert aspects of hegemonic masculinity for computer users. Adult games with sexual content that offer sexualized or objectified images of women could further work to define computers as a technology of decidedly heterosexual masculinity. Stories circulating of men playing *Leisure Suit Larry* or other adult games at work suggest that such software titles may have functioned to exclude or alienate women and foster spaces of homosocial bonding in the office.[23] Nonetheless, adult games did more than simply reinforce computer technology as the purview of white hegemonic heteromasculinity. As these games were developed to be used in the home, software makers also modulated their representations of heterosexuality and masculinity to accommodate their games to the companionate family context in which they were played.

Defining Adult Games

As computers became more widely available in private homes, adult games offered a resource to work through changing norms of masculinity and gender relations. They also served as a potential venue through which to shape the gendered associations of computing culture by representing seduction and dating in the form of a computer game for men to play. As software companies developed these titles, though, they attempted to negotiate potentially incendiary associations with pornography and other adult media. Was the interactivity of adult computer games going to be an intensification of pornography's perceived tendency to offer control over women's bodies? Or would software companies approach sexual material in a manner that contrasted with other pornographic media? Publishers managed this tension both in the representations and mechanics of their games but also in their framing of these games in relation to, or more often distanced from, pornography and other adult media industries.

Computers entered American homes in the 1980s at a time when the relationship between domestic space and sexualized media was undergoing major changes. In part, this renegotiation centered on another home screen—the television. Broadcast television had experimented with representations of the new sexual revolution culture since the 1970s. Networks began airing sexually titillating programs like *Charlie's Angels* (ABC, 1976–81) and *Three's Company* (ABC, 1977–84), but there were restrictions on how much network TV would engage with explicit sexuality.[24] The rise of cable and home video in the 1980s resulted in a new relationship between sexually explicit or pornographic content and the private home screen. In 1986, there were more than one hundred million adult video rentals in the United States and 1,500 different adult video titles available for home viewing.[25] At its peak in the 1980s, the Playboy Channel also made adult content available to eight hundred thousand home cable subscribers, and sexually explicit content was available on dozens of other cable and pay-per-view channels. As Luke Stadel argues, this period constituted "television's reinvention as a sexual technology."[26]

No longer exclusively associated with urban public spaces and adult theaters, the spread of pornography into the home was met with a fervent counterreaction in the 1980s. Scholars describe this period as one in which

American culture was in the grips of a sex panic. In his study of adult video, Peter Alilunas argues that this was in part a response to "the apparent creep of pornography out of traditional adults-only outlets and into retail environments that otherwise catered to 'normal' patrons and families."[27] The ire of this sex panic was directed to adult media industries, and especially adult video stores.[28] The increasing availability of pornography in suburban homes was a controversial development that threatened to evoke protests against companies that helped facilitate this trend.

These contemporaneous sex panics and antipornography efforts influenced the strategies taken by the computer software industry as it sought to target adult users with more mature or sexually explicit content. In developing the adult games market, software publishers negotiated associations with other adult media industries and pornography in different ways. Multiple attempts were made throughout the decade to imagine what an adult computer or video game might look like.

In 1981, Sierra On-Line, which would develop *Leisure Suit Larry* later in the decade, published the adult game *Softporn Adventure. Softporn* was a text adventure created by Chuck Benton, who briefly published it under his own label Blue Sky Software before licensing the game to Sierra (called On-Line Systems at that time).[29] *Softporn* was a fully text-based adventure game with no graphic images. Players explore a nightlife setting in pursuit of sex with various women, which they can achieve by solving narrative puzzles.[30] Beyond the significance of its commercial spread, Laine Nooney has discussed the reception of *Softporn* by the editors and readers of the computer magazine *Softalk* in the early 1980s. She argues that discussions of the relationship between computers and adult content that took place in the magazine sparked significant debates over the role of computing in relation to gender, sexuality, and obscenity. Such discussions sought to determine the social role of the new microcomputing technology. As she documents, the reactions to *Softporn* were mixed—some saw it as harmless, while others saw it as making computers more unfriendly to women and children. In fact, Nooney argues that *Softporn* was more significant for its role in instigating this debate about the place of gender, sex, and adult content in games than as a game in and of itself.[31]

Whereas *Softporn* generated debate about computer culture and its accessibility to different users, the adjacent industry of console video games also experienced its own controversy over sex-themed material with the 1982 release of *Custer's Revenge,* a crude game created for the Atari VCS home console. In this game, players use a joystick to direct a nude and erect General Custer through a barrage of arrows, with the objective to reach a nude Native American woman tied to a cactus on the other edge of the screen and rape her for points. Matthew Thomas Payne and Peter Alilunas argue that *Custer's Revenge* was an attempt to bridge the adult film and interactive entertainment industries. The game was produced by American Multiple Industries (AMI), which began as a manufacturer of plastic storage cases for video tapes before deciding to market adult video games. AMI also licensed the label "Swedish Erotica" from an adult video production company and used this to market *Custer's Revenge* to cement its connection to adult video.[32] Payne and Alilunas argue that the adult video-themed marketing for this game, paired with its poor mechanics, contributed to protests and ultimately led to its failure.[33] This affiliation with adult video was especially contentious in the medium of video games, which had a stronger association with younger boys than even computer games.

As adult games were being developed throughout the 1980s, pornography was associated strongly in the public with a kind of uncontrollable masculine aggression. Antipornography feminists were concerned that pornographic images objectified women and trained men to think of women's bodies as always available for exploitation toward the fulfillment of male pleasure—as objects available for men to control. Pornography, therefore, was being framed as a tool that abetted aggressive forms of masculine sexuality based on domination.[34] Critics were concerned that the computer's capacity to make pornography interactive would exacerbate this style of violent masculine sexuality. This perspective was invoked in protests against *Custer's Revenge*. Those opposing the game, including representatives from NOW and Women Against Pornography, claimed that it expressed that rape was a legitimate form of entertainment.[35]

Later in the decade, a somewhat more developed software industry attempted to revisit adult games. Compared to earlier in the decade, there was a greater concern for how adult games would fit into a computer market that had expanded beyond the hobbyists that made up the audience for *Softporn*. Infocom's adult game *Leather Goddesses of Phobos* (1986) was released the year before *Leisure Suit Larry*. Like *Leisure Suit Larry*, this game relied on humor in its depictions of sexual scenarios; it seemed to anticipate critique by allowing players to choose to play the game in more or less tame or lewd modes.

Other adult games in the 1980s offered explicit images of nude bodies, mostly women but sometimes men, that could be accessed through specific types of play. For example, *Defender of the Crown* (Cinemaware, 1986) featured typical adventure themes and medieval settings that diverged into adult content at the game's conclusion when the player character is rewarded for his success with a quick glimpse of a princess's carefully rendered breasts. There were also many "strip" games released for personal computers. These programs, such as *Centerfold Squares* (Artworx, 1988), simulated board or card games like poker and reversi, but as players win a round the image of naked women, or men in some versions, would be slowly revealed. These games offered a literal equation between successful gameplay and the reward of visual access to women's bodies.

Another adult computer game released around the same times as *Leisure Suit Larry* and *Romantic Encounters* seemed to embody the worst fears of antipornography feminists, heightening concerns that interactive media would exacerbate the perceived negative elements of pornography and offer men a virtual experience of sexually dominating masculinity. This game, *MacPlaymate,* offers an explicit image of a computer-generated woman, Maxie, who lies on her back with legs spread in a pose that offers maximum visibility for the virtual camera and the player. The left side of the screen displays images of sex toys that can be used to attempt to stimulate Maxie's orgasm. Players click on these accessories and drag them across the screen to penetrate and arouse the character. Players are also given the option to change her attire, choosing between complete nudity or a variety of S/M costumes including corset, ball gag, and handcuffs. *MacPlaymate* requires some effort from the player to stimulate Maxie, but the process is not presented as an interpersonal negotiation. Instead, sex in this game is depicted as a mechanical challenge and Maxie is largely yielding. She makes herself fully sexually available to the player who is in control of dressing her and manipulating her body at their will.

MacPlaymate seemed to offer a fantasy of sexually competent and controlling masculinity for men to experience through their control of the game's disembodied avatar. Responding to *MacPlaymate*, critics suggested that the participatory quality of the computer game "takes exploitation light-years beyond X-rated videos and nude pinups," and some even referred to it as *MacRape*.[36] Critics felt that the game, though lacking the resolution of a photographic image, exemplified the potential for interactive media to offer men greater control in their objectification of women. One reviewer argued, "*MacPlaymate* is not pornography as we generally know it. Its users are not passive observers of film or photos.

They are active perpetrators, or worse, interactive perpetrators of brutal sexual violence."[37]

MacPlaymate exemplified fears about computing and its relationship to masculinity, but this was not representative of the way mainstream software publishers approached adult games. *MacPlaymate* largely thrived in bootleg form. It was disavowed by Apple and MacroMind, the company responsible for the animation technology the game's creator Mike Saenz used to make it.[38] The style of interactive sex game in *MacPlaymate* with graphic nudity and easy access to sexualized images of women was in fact unusual at the time. Instead, mainstream software publishers, avoiding explicit imagery in their games, approached the computer mediation of sex with greater ambivalence and more self-conscious depictions of masculinity.

Leisure Suit Larry: "How Not to Meet Women"

Leisure Suit Larry and the promotional efforts around the game were influenced by these ongoing attempts to define the relationship between computers and adult content. Sierra presented *Leisure Suit Larry* as a comic take on masculinity and dating. As described by John Williams, the marketing director for Sierra, *Leisure Suit Larry* was "an adventure game parody on singles life in the 80s."[39] The game largely eschewed explicit sexual imagery, focusing instead on cultivating an experience of cringe that derives uncomfortable humor from the player character Larry's challenges as he attempts to seduce and have sex with multiple women.

The game's focus on humor and development of a cringe aesthetic in its sexual portrayals functioned strategically in a few ways. It allowed Sierra and the designer of the game, Al Lowe, to distance the title from pornographic media and frame the game as a critique of masculinity and sexism. If the fear about computers and sex at the time was that adult games might indulge the desires of men who were looking for control over women's bodies while avoiding complex relationships—similar to the way some critics viewed pornography—then Larry's cringe-inducing exploits and abject failures would help alleviate these concerns.

Although the emphasis on *Leisure Suit Larry*'s humor functioned strategically—allowing Sierra to claim that potentially offensive aspects of the game were in fact critiques of pornography or masculinity—the game was nonetheless shaped by this anticipated scrutiny. Larry did not represent a competent masculinity often associated with a pornographic imaginary. The game cultivated an experience of cringe for players rather than offering an ego ideal through which to vicariously experience sexual competence or power over women. Scholars have addressed the ambiguous depiction of masculinity in *Leisure Suit Larry*, but existing analysis overlooks how Larry's masculinity was framed in comparison to pornographic masculinity as part of an effort to fit the game into the home and companionate family. These accounts also do not consider how the cringe aesthetics used in the game create a more complicated relationship to contemporary masculinity for players, one that commented specifically on Larry's mode of engaging with women in comparison to shifting norms.[40]

Although not necessarily making a radical or feminist critique, through the cringe aesthetic *Leisure Suit Larry* encourages players to maintain some critical distance from Larry's performance of masculinity. The game fosters an ambivalent experience for players in which they are encouraged to both disavow and relate to different versions of masculinity, even if this is pursued to ultimately naturalize a masculinity associated with stable or seemingly rational monogamous heterosexuality.

Prior to its release, Sierra made some effort to frame interpretations of *Leisure Suit Larry* as distinct from pornographic media and as responsive to potential feminist or women's critiques. In a three-part editorial for the magazine *Computer Gaming World* that ran ahead of and then coincided with the release of *Leisure Suit Larry*, the company's marketing director surveyed the state of the adult games market and laid out Sierra's own approach. These articles provide a valuable insight into the way Sierra, and the software industry more broadly, framed its approach to adult games in relation to the history of gaming, to other industries that traded in sexually explicit and adult content, and to public sentiment about computing and its changing social role.

Addressing a computer market that had expanded and shifted since the release of *Softporn*, Williams asserted that Sierra's approach to adult games in the late 1980s did not qualify as pornography. Adult software, as sold by Sierra and other mainstream software publishers, would avoid the type of explicit sex that would earn an X rating if it appeared in film. Instead, adult games, like *Leisure Suit Larry*, would provide the software equivalent of R-rated films, he claimed.[41] Williams also attempted to distance Sierra's games from pornography by suggesting that sex was only one of the adult themes its games would address. For this purpose, Williams named another Sierra title, *Police Quest* (1987), ostensibly aimed at adults for its engagement with violence, drugs, and prostitution, and compared it to the television program *Hill Street Blues* (NBC, 1981–87)—a series famously associated with quality discourses on network television.[42] In making this comparison, Williams suggests that his company's games are adult not because they include explicit sex scenes, but rather because they engage material that would be recognizable or relevant to adult players.

Although Williams used inflated promotional rhetoric, his editorials provide insight into Sierra's understanding of potential critiques of adult games and the market of computer users they would reach. The editorials suggest that Sierra believed it could not just sell *Leisure Suit Larry* to isolated men or computer geeks, but rather had to imagine how its products would fit into companionate family relationships. Williams claimed that Sierra and other software publishers making adult games were not selling to "the traditional entertainment software market (i.e., teenage boys)" but rather to adults.[43]

Still, Sierra had to acknowledge that in a domestic family context, games for adults could be accessed by children and teens as well. As such, it used a trivia game to screen potential players—access to the game was granted only to players who got no more than one answer incorrect. The trivia game covered topics related to political and popular culture references that would likely be familiar only to adults. Lowe admits that Sierra didn't necessarily believe this would deter young players—in fact, there was a trick that allowed users to bypass the quiz by depressing the Alt and X keys—but it speaks to Sierra's consideration of the domestic and family context for these games. Even if they would be played by adult family men (as reviewers of the game later made a point to label themselves), these men were part of specific domestic contexts and relationships. Sierra felt that *Leisure Suit Larry* needed to be posed in such a manner to assuage anxieties that it would negatively affect these relationships.

Within Williams's discussion of anticipated critique, the reaction from women seemed a prominent concern. In contrast to what he and many others characterize as the violent, aggressive, and misogynistic sexuality in games like *Custer's Revenge* (which he called out in his editorial), Williams argued that Sierra's *Leisure Suit Larry* was a socially responsible game. In his last article in the series, published after the game was out, Williams even claimed that many women wrote to Sierra expressing their approval of the game for exposing macho men as jerks. Williams concluded that *Leisure Suit Larry* critiqued chauvinism and would help men who were not yet "enlightened toward women."[44]

Sierra's comments about the game's treatment of masculinity suggests concerns that it might be perceived as antifeminist or sexist by contributing to the objectification of women. Of course, it is important not to make the mistake of taking Williams's claims as wholly sincere statements about the publisher's prosocial intentions. For example, Williams claims that *Leisure Suit Larry* is "socially conscious" and cites a company spokesperson who said it will "advocate 'safe sex.'"[45] What this translates to in the game is that if Larry sleeps with the sex worker character without wearing a condom, his crotch catches fire and he drops dead. This is clearly not a serious or high-minded treatment of sexual health. Still, Williams's claim anticipated critiques about the potential dangers to women of pornographic or adult software.

Williams's discussion shows how women, even if not the expected primary audience that would play *Leisure Suit Larry*, were considered in its design and promotion. His editorials referred to Tipper Gore's battle with the music industry as a cautionary example for software companies. He also related a story about a mother who was scandalized when she saw another adult game, *Leather Goddesses of Phobos*, being advertised at her local computer software retailer, fearing it would corrupt her son.[46] Granted, from Williams's perspective, women were of concern only as an oversensitive maternal watchdog the company had to navigate around.

The game's creator presented consideration of women as part of the audience for his game differently. When writing jokes and puzzles for *Leisure Suit Larry*, Lowe claims that he used his wife—whom he called "a sensitive and modern, liberated woman"—as a guide to know if a joke was appropriate. As he explains, "If she thought it was funny, it was funny. But if she found it too raunchy, it was too raunchy."[47] Lowe's story, too, suggests that there was a concern about women's potential reactions to *Leisure Suit Larry*, but he seems to offer a gentler image of women as reasonable moderators of good taste rather than as irrational or overly sensitive critics.

As Lowe's account indicates, concerns about reactions to *Leisure Suit Larry* from consumers, including women and families, shaped the game itself. Although Sierra's claim that *Leisure Suit Larry* was not pornographic was part of its promotional efforts to avoid associations with more controversial adult media, the game did in fact avoid explicit depictions of sex. This was not a game that re-created a pornographic imaginary of masculine dominance by granting players power to access and manipulate images of women and their bodies. Instead, *Leisure Suit Larry* focused on creating an experience of cringe reactions to masculinity and represented sexual scenarios fraught with anxiety.

Experiences of cringe in *Leisure Suit Larry* often arise from the character's ineptitude at relating successfully with women. As Havas and Sulimma theorize its use in dramedy television, cringe is not only concerned with comedy about bodies, but also depicts transgressions or failure to meet social norms and expectations. Among other qualities, cringe protagonists often navigate embarrassing situations "while failing at communication, exhibiting unawareness of expected social behaviors, and having their self-images diverge from the ways others perceive them." As Havas and Sulimma explain, the viewer affect encouraged by cringe aesthetics is ambivalent and conflicted. It often can oscillate between relative distance or recognition from the awkward situation on screen, but it also involves a "shiver of embarrassment, awkwardness, disgust" that creates discomfort for the viewer nonetheless.[48]

In the mobilization of cringe in *Leisure Suit Larry*, too, this conflicted position and uncomfortable affective experience is also present. In moments of cringe, players are not meant to identify exactly with Larry, but their distancing from the character is neither complete nor comfortable. Instead, the player is put in a position to experience a shiver of embarrassment or awkwardness both in moments of recognition of Larry's social transgressions and in moments when the player establishes distance. *Leisure Suit Larry* fosters a highly ambivalent play experience that has the potential to encourage reflection and critique of masculine norms and gender relations.

Despite its sexual narrative, experiences of cringe are pervasive in *Leisure Suit Larry*. The player explores a satirical version of Las Vegas (called Lost Wages) with the goal of directing Larry to lose his virginity before the night is over. *Leisure Suit Larry* was a graphic adventure game remake of Sierra's earlier *Softporn* and borrowed from its predecessor many of the same adventure game puzzles to drive forward the progression of play. The game allows Larry to pursue three primary women characters: an unnamed sex worker; Fawn, a woman the player meets at a disco and who robs him after a quickie wedding; and Eve, a final dream woman whom the player meets in her penthouse hot tub.

Initially, *Leisure Suit Larry* may appear as a straightforward adventure game that translates the seduction of women into gaming challenges that can be solved through good play. In this interpretation, the game would seem to offer access to a pornographic imaginary in which access to women comes from skillful performances of masculinity at the keyboard.

For example, to have sex with Eve at the game's conclusion, the player must have collected an apple elsewhere in the game to offer her. If he has failed to do so, progress will be blocked, and the player must return to earlier game locations to proceed. Figuring out puzzles like this to win the game results from familiarity with genre conventions and clever puzzle-solving skills. Discussing sex scenes and sexual consent mechanics in more recent mainstream video games, Josef Nguyen and Bo Ruberg argue that in many games with sexual material, consent and sex are treated as a reward to be earned through performances of skill and effective play.[49] Although they are discussing more contemporary games, this pattern, to some degree, is present in *Leisure Suit Larry* as well.

Yet, although the game treats sex as a reward for good play in the overall narrative and game progression, it presents this sexual material through a cringe aesthetic that does not present a triumphant masculinity. For example, if the player chooses to "talk to" the sex worker after sleeping with her, Larry asks, "Was it good for you?" She responds witheringly by asking, "Was what good?" A moment like this is an example of cringe in which the player may feel embarrassment for Larry's misunderstanding of the sexual exchange that has just occurred. Here, the player likely knows better and can judge this slight from a distance.

In other moments, the game brings the player closer to Larry's position. For example, though solving puzzles throughout the game may provide the player a sense of achievement, the game refuses to reward the player with access to sexual imagery. Two of three sex scenes in *Leisure Suit Larry* proceed behind a black censor bar, while the other takes place off screen entirely while the game cuts to fireworks (Figure 12). By denying this visual access, the game appears to bring the player closer to Larry's own position of humiliation and embarrassment. In moments where players expect to have unlocked a sex scene but are thwarted, they, too, are the butt of the joke—not unlike Larry, who thinks he is going to have sex with Fawn at one point only to be tied up and robbed.

Leisure Suit Larry reinforces an ambivalent and cringe-inducing relationship to the philandering masculinity embodied by Larry throughout the game and in accompanying promotional material. Larry is often presented as an unappealing avatar of masculinity whose relationship to women is not ideal. The instruction manual for the game introduces Larry as a software salesman approaching forty years old and still living with his mother. He is an aging virgin with a "receding hairline and expanding waist."[50] In Larry Laffer, the figure of a playboy who puts off marriage to pursue his own pleasures is turned into a nightmarish image.

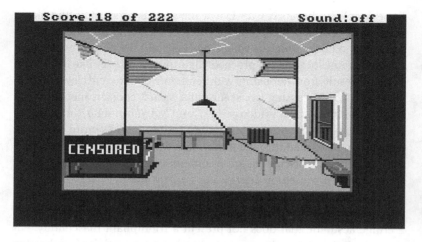

CENSORED

Figure 12. Screenshot from an emulation of *Leisure Suit Larry*. Images of sexual activity are obscured by a censor bar.

Lowe emphasizes that he intended Larry's performance of masculinity to appear ridiculous. He claims that when Sierra president Ken Williams asked him in 1987 to program a graphic adaptation of *Softporn*, the original game had become too dated. Rather than update *Softporn*'s references to account for these perceived changes in singles culture, Lowe took a humorous approach that emphasized the datedness partly through his invention of the character of Larry, who hadn't appeared in *Softporn*.[51] Lowe does not make clear exactly what about the earlier version of masculinity appeared dated to him. But the exaggeration of Larry's leisure suit and his performative pursuit of a kind of swinging singles life suggests that Lowe perhaps found a more "traditional" monogamous masculinity (like his own) to be better attuned to the 1980s. During gameplay, characters often respond negatively to Larry; some call him a dork and a creep and frequently comment on his toxic bad breath. In addition to his out-of-fashion attire, Larry's pickup lines are outmoded and the women he talks to seem to instinctively try to get rid of him—at least until they find ways to exploit him. As such, Larry appears to offer a version of cringe-inducing masculinity that players could distance themselves from rather than interpret as an ego ideal.

This representation of derided masculinity in adult media targeted for men is not unprecedented. In fact, in this respect *Leisure Suit Larry* is not completely divergent from some strains of the pornographic media from which Sierra attempted to distance its game. Constance Penley argues that humor plays a significant role in porn, often at the expense of men— a "bawdy populist humor whose subject matter was so often the follies and foibles of masculinity." Like this longer history of bawdy humor, *Leisure Suit Larry* revels in gross and scatological jokes at the expense of Larry's bodily integrity. In the course of the game, Larry gets peed on by a dog, in some endings gets caught attempting bestiality with that same dog, falls into a garbage dump, and drowns in a broken overflowing toilet in a seedy dive bar. The abject is also reflected through the game's setting and aesthetics—a dilapidated watering hole, a filthy bathroom, and a depressing chapel. These spaces are depicted in a cartoonish style, but visual details like cracked ceilings, crumbling walls, and trash-littered alleys create an impression of a sleazy singles scene.

The game's representation of women characters mirrors Penley's description of "trickster women with a hearty appetite for sex" that populate stag films and other pornographic media, where women are the ones "both initiating sex and setting the terms for the sexual encounter."[52] The character Fawn, for example, maneuvers her encounter with Larry to extract jewelry, candy, flowers, money, and a marriage before agreeing to sleep with him. Once in the honeymoon suite, it is hinted that Fawn has a satisfying sexual fling with the man delivering the alcohol she has sent Larry out to procure, and then she ties Larry up and robs him before exiting the game. This representation of women is not necessarily feminist. As Adrienne Shaw points out in her discussion of *Leisure Suit Larry*, the game takes shots at women's liberation; Lowe's claim that the game is feminist because women get the upper hand is a "*profound* misunderstanding of feminist politics."[53]

Still, these elements of *Leisure Suit Larry* point to a mode of representation of sex that is not based only in domination or an imaginary of masculine sexual competence. Instead, Larry's relations with women prompt cringing due to his misreadings of social cues. Contrary to some antipornography feminist critiques, representations of masculinity in porn, according to Penley's account, can be more critical of—or at least seem to attempt to work out anxieties about—changing cultural norms. Similarly, *Leisure Suit Larry*'s focus on humor allowed it to fit into a kind of men's cultural tradition that also featured critical depictions of masculinity.

Although *Leisure Suit Larry* provides a cringe-inducing depiction of the failures of masculinity that players could potentially disavow, at times the game also encouraged identification with Larry. *Leisure Suit Larry*, in fact, often focuses its cringe aesthetics on moments of social failure in which the player is implicated. In this case, identification does not equate simply to sympathy for Larry and his pursuit of women but holds the potential to involve the player in moments of embarrassment as well.

Leisure Suit Larry is presented predominantly through a third-person camera perspective. To navigate the game, players use their keyboard arrow keys to direct Larry to walk around. Other actions require players to type a command like "open door," "take candy," or "talk to girl." To access more detailed images and talk to women in *Leisure Suit Larry*, players must first establish eye contact by typing a version of the command "look at girl." The resulting close-ups provide the most graphically detailed character images in the game. These moments are also the instances in the game when the player's visual perspective aligns most closely with Larry's.

This aspect of the game is even included in Sierra's promotional material that announces an exciting feature of *Leisure Suit Larry*: "Full facial view of game characters. As Larry tries out his best pick-up lines, you can watch as the girl changes her facial expressions. If she smiles or winks, Larry may be in for the time of his life."[54] As described in the manual, these appear to be moments of visual pleasure, but these views are also saddled with anxiety for the player. Once the player establishes eye contact, he is quite limited in the inputs he can type to engage with the woman. The player can type "talk to woman," but then Larry proceeds with his preprogrammed pickup lines asking the character such clichéd questions as what her "sign" is and calling her "baby" and "sweetheart."

These are important moments in which the game encourages the conflicted experience of cringe. Aligned with Larry's point of view, the player watches the woman's expression change as Larry humiliates himself (Figure 13). When playing *Leisure Suit Larry*, the user is subjected to the awkward attempts Larry makes to impress women. The player is brought closer, too, as he monitors these characters' facial expressions to get a sense of whether he has navigated the game's puzzles correctly. Not only do these moments task visual pleasure with the pain of rejection, but they also refuse to provide unadulterated triumph. Even if they have mastered all other game challenges, players cannot use better lines when talking to women and must look on from Larry's point of view as these women regard him with utter contempt. As one reviewer puts it, "No matter how well you play the game, he remains a jerk."[55]

Lowe's description of the origins of the Larry character also speaks to the way the cringe ambivalence in the game functions. In order to delimit the bounds of acceptable masculinity, cringe aesthetics assert distance at times and admit similarity at others. Lowe, a computer programmer and game designer, claimed to model the character of Larry off a software salesman who sometimes visited Sierra's offices and was disliked by Lowe and the other programmers. This suggests a context in which computer workers attempt to make fine distinctions to establish a hierarchy of technologically adept masculinity.

At other times, Lowe intimates that there are similarities between himself and Larry. Lowe calls himself "the world's oldest computer games

Figure 13. Screenshots from an emulation of *Leisure Suit Larry* show how Faith's reaction changes in response to Larry's more or less successful pickup lines.

designer." When the game was made, he was actually older than the character of Larry, who was ridiculed for getting on in years. Granted, unlike Larry, Lowe was a married man and therefore distanced from Larry's brand of swinging-singles masculinity. Still, in his letter that accompanied a later reissue of the game Lowe makes self-deprecating comparisons between himself and Larry. For example, Lowe informs the reader, "Hair has always been one of Larry's downfalls (pun intended!). Follow the numerous references throughout all the games to hair loss, wigs, barbers and baldness. Carefully study my photo on the back of the box. Draw your own conclusions."[56] Lowe thus implies that some of Larry's ostensibly negative attributes, the ones the game ridicules him for, are shared by his creator. Lowe's discussion of Larry shows how the character can function to inspire reflection about one's own masculinity even if the player also ridicules Larry.

Appraisals of *Leisure Suit Larry* by contemporary game scholars and computer historians have acknowledged its ambivalent approach to masculine representation. Computer historian Jimmy Maher points to the fact that the game not only ridicules Larry, but also offers denigrating portrayals of the immigrants and people of color Larry encounters throughout his adventure. The only Black character in the game is a pimp whom the player can distract from his duties by turning a nearby television set to a porn channel. And in a scene in a convenience store, the game relies on humor at the expense of the store clerk's exaggerated fictional accent. As a result, Maher concludes of the game, "It's all about mean-spiritedly making fun of people who are—or who are perceived as—weaker and more pathetic than the people who made the game."[57]

In a brief essay primarily addressing LGBTQ representation in the *Leisure Suit Larry* series, Shaw also offers some insights about the game's treatment of heterosexual masculinity. She argues that despite the diversity of the players who may play the game, "The jokes being told or shown, imply a player who has a similar identity to Larry (i.e., a heterosexual, cisgendered male)"; in her estimation, this is a game "where heterosexual masculinity is the goal."[58] Although Shaw finds some potential lessons in the game on which better LGBTQ representation could be developed, she argues ultimately that the series fails in its "attempts to use humor to undercut its own oppressive messages." According to Shaw, the player is meant to be on Larry's side: "Heterosexual masculinity is joked about in the games, but it is not The Joke."[59]

I agree with Shaw and Maher about the game's depictions of queer characters and characters of color—in these cases, it does not remark on or denaturalize the whiteness or heterosexuality of the player as it does in its depictions of Larry. Yet, it is not entirely accurate to say heterosexuality is the goal in *Leisure Suit Larry*; the game works to distinguish between different types of heteromasculine performance and offers a critique of the swinging sexuality represented by Larry. It is more precise to point to the ways that *Leisure Suit Larry* helps work through the limits and expectations for masculinity. It does so by using cringe and creating an ambivalent position for players with respect to Larry.

It is common for scholars who bring up *Leisure Suit Larry* to situate it ultimately as an endorsement of heteromasculinity. In their discussion of *Plundered Hearts* (Infocom, 1987), another contemporaneous game about romance and sexuality, Anastasia Salter suggests that a heteronormative male gaze is encoded in *Leisure Suit Larry*. They point to the many references to women's body parts and focus on sexual acts in the game's source code as evidence of this orientation in the game. They conclude, "The integration of Larry's pursuit of scores . . . on a point system is particularly reductive, and while his attempts at conquest are frequently thwarted, his persistence is continually cast as a virtue."[60] It is not Salter's goal in the essay to provide an extended discussion of this game; instead they provide a nuanced analysis of *Plundered Hearts*.

In fact, Salter contends that the prominent place that *Leisure Suit Larry* holds in game scholarship adds to the centering of a masculine and heteronormative perspective in game studies.[61] In my own analysis, though, I discuss this game to complicate this masculine perspective. Tracing the history of *Leisure Suit Larry* shows ways that anxieties about women's concerns and critiques of masculinity influenced the game. Furthermore, understanding how the game functions and positions players helps better illuminate the way that it endorses some versions of hegemonic masculinity over others and leaves room for contestations or defamiliarizations of contemporary norms.

Reviews of *Leisure Suit Larry* demonstrate how some players worked through the game's conflicted cringe representations and fluctuating identification with Larry's performance of masculinity. Some reviewers exhibited a self-conscious attitude when discussing their position in relation to the game and frequently commented on the complex relationship between the player and the game's protagonist. As if to signal their own more rational and sensitive masculinity, reviewers often found it necessary to disclose their marital status or awareness of feminist critiques of masculinity before evaluating the game. But this distancing from Larry was not complete.

Before his positive review in *Computer Gaming World,* Roy Wagner offers a disclaimer assuring readers that he is a "wholesome family man," unlike Larry himself, but that he could nonetheless enjoy the game's humor. He later explains, "Half the fun is relating (vicariously or not) to what is going on as you travel by cab from Lefty's Bar to the Casino, Chapel, and Store."[62] Reviewers also frequently proclaimed their judgment of the game as sexist while still expressing pleasure in identifying with Larry. Before describing the game as "funny and fun to play," one reviewer acknowledges "many women will probably be incensed" and notes "it will probably take a completely separate game to remove the male bias from *Leisure Suit Larry*."[63] These reviewers labor to distance themselves from Larry even as they express enjoyment and pleasure in the game. Their disclaimers demonstrate that the representation of masculinity in the game encouraged players to consider their own masculinity in comparison. Even for reviewers who enjoyed the title, their reviews show that they were forced to consider what such pleasure meant in relation to contemporary heterosexual relationships and gender roles.

Admittedly, the game's choice to depict a "dated" masculinity rather than to engage directly with the New Man of the 1980s had some potential to undercut the game's critique. In a piece titled "How Not to Meet Women," one reviewer expresses disappointment in the "many examples of a fifties mentality" and says that despite the "traditional-male laughs" the game could have been funnier.[64] Another reviewer comments on the game's "unsophisticated view of sex, culled from the philosophy of Hugh Hefner, circa 1962."[65] By noting this dated masculinity, reviewers could critique the attitudes exemplified by Larry without necessarily implicating their own views or treatment of women. In fact, reviewers appear to use the game's cringe aesthetics to disavow what they see as the pathetic excesses of sexual liberation as it is reflected in a kind of nonmonogamous 1970s masculinity. Instead, they reasserted a masculinity more like their own stated position—a heterosexuality authorized by marriage and membership in a professional middle class. Still, even these reviews suggest that those playing and reviewing the game felt the representation of Larry's masculinity required comparison to their own, even if only to deny the comparison.

Although *Leisure Suit Larry* orchestrates a complex relationship between the player and Larry, the place where it undercuts this critique to the greatest degree appears not through the character himself, but rather through the masculinity represented by the narrator and the many comedic exchanges it facilitates for players. In *Leisure Suit Larry,* players experience a running commentary as they direct Larry through his encounters. These create moments of cringe in which the player is given an avatar of stable masculinity to identify with in judgment of Larry.

The computer is presented as a rational masculine interlocutor accompanying the player in his judgment of Larry. For example, when the player directs Larry to look in a bathroom mirror, a text box pops up: "You see a handsome, dashing, macho, sexy, young man. (Obviously, this mirror has quite a bit of distortion!)" The parenthetical aside is not Larry's commentary, nor the player's, but represents a seemingly objective voice—one that appears to represent the game or the computer itself. This commentary often pops up at moments when the player has made a fatal error or acted in a way that seems to get Larry in trouble. After Larry gets mugged, the voice says, "Larry, when are you gonna learn to stay out of those dark alleys!!" Or when Larry is castrated and dies after attempting oral sex, it says, "Sorry, Larry. No oral sex in this game. Suffer!" Or when Larry loans Fawn money and agrees to marry her after they have only met a few minutes prior, it asks, "(What are you into this time, Larry?)" While commenting on Larry's behavior, these moments indicate a masculine voice of reason that the player can identify with as an ideal, rational masculine perspective. Despite critiquing one version of masculinity, through Larry, in which the player is potentially implicated, the game provides a more objective ideal of masculine rationality in the disembodied voice of the computer that comments on Larry's actions.

In fact, in an interview, Lowe discussed this element of the game: "People always ask, 'Are you Larry?' I'm not, although I am in the games. I'm the narrator. When Larry asks some dumb question and the narrator responds with some smart ass remark, that's me."[66] This relationship between the player and a more objective narrator or authorial voice may undermine the game's critique of masculinity. Whereas the game might at times use cringe to incite players to consider their complicity or similarity to the failed version of masculinity on display, it also offers this more stable point of identification at other times—a rational voice largely distanced from the same failures, frustrations, and difficulties that Larry exhibits.

Romantic Encounters at the Dome: An Adult Text Experience

Romantic Encounters at the Dome, released the same year as *Leisure Suit Larry,* takes a different approach to critiquing masculinity and interpersonal, sexual relations. *Romantic Encounters* also relies on cringe aesthetics to produce an ambivalent and shifting relationship to the player character in the game. Yet, whereas *Leisure Suit Larry* develops cringe-inducing experiences by exaggerating Larry's dated masculinity and the equally abject nightlife settings in the game, *Romantic Encounters* represents its main character as a sensitive New Man contending directly with contemporary norms of heterosexual masculinity and shows his shortcomings in relating with women in a more aspirational context. This is a representation of cringe that is harder to disavow as categorically different from the player's own masculinity. *Romantic Encounters* develops its player character as a seemingly rational figure who appears to successfully embody a contemporary masculine ideal. But this rationality is ultimately undermined in moments of cringe in which the character's instrumental attitude toward women becomes apparent. Like feminist critics of the 1980s new masculinity, the game helps show how sensitive masculinity can be pursued in ways that are more self-serving than feminist.

Romantic Encounters was created by Lee Thomas and published by Microillusions, a small San Fernando Valley company that specialized in software for the Commodore Amiga. Unlike the popularity and profusion of discourses about *Leisure Suit Larry, Romantic Encounters* appears to have been a more obscure game even when it was released. It was marketed as part of Microillusions' "adult games" category, which also included the gambling games *Black Jack Academy* and *Craps Academy.*[67] Like Sierra, Microillusions used the label of "adult games" to refer broadly to themes and activities that could be of particular interest to adults and did not only refer to sexually explicit content.

Romantic Encounters is set in a private Los Angeles nightclub called the Dome, where the player is given the ability to interact with a cast of characters to explore social and sexual connections. The game offers the option to select to play as "male" or "female," which determines the gender of the characters the player can interact with romantically. It assumes a heterosexual orientation for players. For players that select to play as "male," the game offers seven primary women with which to interact. All are described as young, slim, and attractive, although the player character comments on their differing body types and degrees of sexual appeal.

For example, while Cathy is described as pretty and feminine with "beautifully modest curves," Kitty is a "shoe-in as playmate of the year" and "a very sexy lady with large breasts and the legs of a dancer." Although it is a text-based game, the software packaging depicts the Dome and its patrons as glamorous and upscale singles, drawing on a vaguely new wave or cyberpunk aesthetic inspired by LA's postmodern architecture (Figure 14). This packaging indicates the more aspirational version of masculinity that is presented and critiqued in *Romantic Encounters*.

Not as much information is available about the production of *Romantic Encounters*. Despite its unique approach to interactive fiction, executed with wit and originality, Microillusions does not appear to have promoted it heavily, and it received limited reviews from critics. According to Veronica

Figure 14. Front cover of the software packaging for *Romantic Encounters at the Dome*, which depicts the game's aspirational setting.

Thomas, who was credited as providing psychological services for the game and was married to its creator, Lee Thomas wrote *Romantic Encounters* largely in isolation and only solicited interest from software publishers later in the development process. In fact, her account of the creation of the game and the role of computers in the Thomas home mimics many of the anxieties about computers' isolating effects circulating at the time. Lee Thomas's purchase of the computer was initially a point of contention in their family because of its cost. Lee, whose computing and electronics skills were self-taught, used the computer to program *Romantic Encounters.* Having failed to find sustained success as a playwright, Lee's time programming the game allowed him an outlet for his rich fantasy life, but it seemed at times to distract from his companionate relationships.

Even as he programmed the game in relative isolation, Veronica Thomas says that Lee would occasionally solicit her opinions on dialogue and scenarios for *Romantic Encounters* as he was developing them. As a psychiatrist herself, she seemed skeptical that her input would be defined as formal psychological services or consultation. Instead, she speculates that it is possible Lee's writing of women characters in the game may have been influenced by his experience working for a matchmaking company, where his job involved interviewing women around Los Angeles about their lives as part of the screening process.[68] The attribution of psychological services for the game may have been spurred by Microillusions or as a strategy to appeal to the publisher.

Despite the limited information about the development of *Romantic Encounters,* Veronica's account suggests some ways in which the game's creation may have been shaped by the companionate context in which Lee wrote the game and his observations of the singles experience in Los Angeles where the game was set. To some degree, Veronica suggests that Lee viewed the game as an outlet for fantasy, evident in its aspirational and glamorous setting, despite the occasional undercutting of its protagonist.

Rather than an easy substitute for relations with women, *Romantic Encounters* directly simulated encounters between singles, reproducing the challenges of these interactions. Whereas *Leisure Suit Larry* offered the possibility of translating seduction into arbitrary game puzzles that required collecting items and using them in the right configuration, the challenges in *Romantic Encounters* were interpersonal. The player was required to figure out what the game considered to be the right decisions and text inputs to progress.

Romantic Encounters may be more accurately described as an inter-active fiction game rather than an adventure game. There are no objects to collect or game puzzles to solve. Players' success depends on their choices at key branching moments in the story. Often these choices are binary. For example, a player can choose whether to accept a woman's invitation to her room. At other times, the range of acceptable inputs and their effects are harder to determine ahead of time. When pursuing a woman named Jeri, for example, if the player accepts an invitation to her apartment in the Dome, the game offers the following prompt: "You try to figure out what to talk to Jeri about, to keep her relaxed, to get closer to her. Now you start the conversation." The player is then required to type in what to say to Jeri. If the player attempts a lewd or aggressive approach, like commenting on her breasts or immediately proposing sex, Jeri recoils and asks him to leave. If the player tries various lines to make small talk, the game responds by noting that it is difficult or uncomfort-able to figure out how to stimulate Jeri's interest. The way to move forward at this point in the game is to remember that Jeri has expressed an inter-est in photography and to ask a question relating to this hobby.

The difficulties and puzzling aspects of interpersonal relationships in *Romantic Encounters* are not displaced onto arbitrary adventure game puzzles. Instead, the player is forced to contend with the complicated nature of interpersonal relationships by crafting lines, figuring out con-versation, and knowing when it is appropriate to initiate or reciprocate sexual contact. *Romantic Encounters* seems to test the skills of listening and being sensitive to others.

The player must also exercise sensitivity in game choices to main-tain successful sexual encounters or risk experiencing cringe-inducing rejection. In *Romantic Encounters,* sexual encounters must be carefully negotiated or they will end prematurely and often in embarrassment— although they often end in embarrassment anyway. For example, after initiating sex with a character named Tanya, the player finds that she becomes overly passive in bed. At this point, the game requires the player to act to move forward. The player can try to shake Tanya out of her trance, at which point she will explain that her past sexual partners have preferred that she act passively in bed but she will readjust by trying to be more engaged. Alternatively, if the player goes with a different approach, Tanya becomes uneasy and confused and the computer reports, "What-ever intimacy you still had begins to drift away." Eventually, Tanya extri-cates herself from bed to hide in the bathroom, prompting the player character to leave. Players are encouraged to be attentive to the reactions of non-player characters to more successfully craft strategic reactions and maintain successful relationships.

Romantic Encounters uses cringe aesthetics more sparingly than does *Leisure Suit Larry*. In *Romantic Encounters*, the player character, initially presented as a more idealized version of sensitive new masculinity, is encouraged to enact this attitude in his interactions with non-player characters. Only as the game progresses are this characters' faults revealed and he begins to invoke moments of cringing reaction.

The player character in *Romantic Encounters* is not irresistible to women. For example, when introduced to Clarissa, a beautiful model, she refuses to acknowledge him. Yet, it is suggested that he is attractive enough that many women at the Dome seem receptive to engaging with him, and some even take the initiative to pursue him. Furthermore, the player character is initially portrayed as a seemingly sensitive man. He reflects on other characters and his surroundings in ways that suggest a perceptive and empathetic nature. In fact, the game depicts the player as having a romantic sensibility, one that can intuit deep emotions in others. For example, in response to one of the hostesses, the player contemplates, "You stare at this hypnotic woman, wonder what she might know about you, what special insight she might possess. In parting her deep eyes seem to suggest another moment in time together. Then she leaves." The game also implies that the player character is genuinely interested in getting to know the women he meets. Describing his interaction with Jeri, the game reports, "You start discussing movies you have both seen, novels you have read, and then when you find yourselves in quick agreement on the parts that excited you, stimulated you, a self-consciousness settles in. You find you are making exciting connections with Jeri and she with you, not only in conversation but on another, deeper, unspoken level."

By representing the player character as a seemingly desirable New Man who is adept at interacting with women, it might appear that *Romantic Encounters* is offering male players a vicarious fantasy of seducing women. Yet, as the game proceeds, this sensitive and idealized masculinity is revealed to be an illusion. By invoking identification with this player character as an ideal but then gradually undermining him, the game makes possible a cringing awareness in players of their own instrumental play.

The player character is shown to view women instrumentally and exposes a sense of entitlement to those he meets. He is also quick to anger when he perceives he is slighted. For example, if after initially hitting it off with Jeri the player is unsuccessful in getting an invite to her apartment and she excuses herself to leave, the player character is indignant. The game gives the player this glimpse into his reaction: "You're angry with Jeri and with yourself. Then you start getting annoyed because you just bought Jeri a four-dollar drink, and you start reflecting how women always get to take a guy for drinks, dinner, a movie, whatever." At moments like these, the player is made aware of the character's embarrassing sense of entitlement with regard to women he is pursuing, the same women the player himself has tried to manipulate through text inputs. *Romantic Encounters* cultivates a conflicted position of identification and intimacy with the player character in these moments of failure. Even as the player must recognize his similarity to the character, the game also invites critique. After Jeri leaves, the player character heads home and accidentally gets urine on his shoes while peeing in the parking garage.

As it proceeds, *Romantic Encounters* exposes the player character's cringe-inducing attitudes toward women and consent, conflating them with the player's own attempts to advance in the game. The character's approach to seduction is strategic and instrumental. In this game, the player learns about the character's views through a text-based internal monologue. For example, when a woman invites the player to her office for casual sex, she begins to discuss her dissatisfaction working at the Dome and the player character observes, "Bobi starts to seethe. You instantly spot this as a danger sign. If you let her get too involved in other things, she's going to start dealing with you in a secondary way. You look around for the champagne she had advertised. This could help get her on the right track." In these moments, the player character shows his awareness of the moods and feelings of his partner, but only insofar as these are cues he can use to improve his chances of having sex with her. Yet, by explaining the character's interactions with women this way, it also indirectly instructs the player on what actions to take to succeed with these women in the game. Like the character, the player now knows that he can use champagne to maintain the interaction with Bobi.

For the player character, sensitivity to women is predominantly a means to an end, a way to keep from having his sexual prospects shut down. The player character's sensitivity is shown to be aligned with the manipulative techniques of the pickup artist. By providing insight into the player character's motivations, the game critiques this version of performative, manipulative sensitive masculinity yet simultaneously aligns it with the player's own strategic gameplay. Unlike *Leisure Suit Larry*, *Romantic Encounters* allows players to hone and develop sensitivity as a game skill. Yet, when players perform sensitivity in the game, this does not change the fact that it is done in an instrumental way to win women over and they cannot entirely escape the game's rebuke.

Romantic Encounters implicates such attitudes toward women even when they are performed as part of what seems like sensitive masculinity. It also undermines the character's supposed masculine rationality. It becomes clear that he is not nearly the thoughtful and reasonable man he appears to be. Nearly all the player character's perceptions and actions are colored by masculine insecurity and nagging sexual desire. For example, after an unsuccessful attempt to win over the sex worker he is attracted to, the character tells himself the following: "With Roxy it was just a matter of bad chemistry, you reason! Another hooker might have fallen madly in love with you! You start feeling better as you start to drive and begin randomly scanning back on many of your everyday but not-so-ordinary accomplishments! Swimming the length of an enormous pool underwater, with one breath! Finding the first girl who ever loved you—rejecting her! Finding the second girl who ever loved you—rejecting HER! Making 5 baskets in a row from the free throw line! The list really begins to add up." Though not a comedy game like *Leisure Suit Larry*, *Romantic Encounters* reserves its humor for these select moments of cringe when the player glimpses his character's internal monologue as he rationalizes his behavior or compensates for his sexual failures.

This humor also comes out in subtle ways to highlight the fallacy of the player character's presumed rationality and logic. For example, if the player tries to pressure Jeri to have sex with him and is rejected, the character convinces himself he is lucky to have flushed out Jeri's "insanity" before getting more involved with her. The game reports, "You take a deep breath, start to relax, experience the satisfaction your logic has produced in you. You start to compliment yourself on having the intelligence to flush out a potentially troublesome relationship . . . for not getting hung up on some crazy woman who could make life difficult for you . . ." Similarly, when Tanya rejects him after he fails to meet her expectations, the player character similarly congratulates himself: "A comforting feeling of sanity comes over you. You start to compliment yourself on being a flexible and mature kind of person. You think about some of your other assets: an ability to keep your feet on the ground when other people are losing their heads. You reason this is a good asset to have, one that will work well for you and your future." The tone of self-satisfaction in these moments undermines the professed logic of the character's rationalizations. Here, the game appears to elicit from players a sense of embarrassment in recognition of their own strategic version of play.

In *Romantic Encounters,* the gradual shift to realizing the weaknesses of the seemingly ideal player character has the potential to disrupt the player's strategic play and inspire reflections on this mode of relating. Furthermore, the critique of masculinity in *Romantic Encounters* has the potential to challenge masculine norms more than *Leisure Suit Larry* does, targeting a masculine figure that initially appears as a rational, seemingly "universal" subject for player identification. Yet, the game reveals that this rational individualism is an illusion. The player character's insecurity colors nearly all his interactions and perceptions. His constant references to his own logic are shown to be desperate rationalizations for his failures. Because the player character appears like an abstract or neutral masculine figure at the start, the critiques of his deficiencies are more difficult for the player to disavow, leaving the player no stable point of identity in the game. Ultimately, *Romantic Encounters* critiques masculine rationality itself, especially as it pertains to an instrumental attitude toward women and sexual relationships.

Adding to the potential that the game could foster reflection on masculine attitudes toward women, *Romantic Encounters* also featured a mode with a woman player character, depicting encounters at the Dome from a different perspective. The characters and relationship arcs featured in the female version of the game largely paralleled the male version. Yet, whereas the major challenge for the male player is to avoid failing at an encounter and having it end in disappointment or embarrassment, the female version depicted these encounters as potentially more dangerous or exploitative.

For example, the female player might encounter Paul, a character that parallels Bobi from the male version. Like Bobi, he works at the Dome and invites the player to an office upstairs where he starts to rant about being underappreciated by his boss. But whereas the male player character is shown to be worried that Bobi will lose interest and the player is encouraged to distract her with champagne, Paul initiates a rapid shift to sexual activities and kisses and gropes the female player character. Throughout the encounter, the female player character's interior monologue expresses that she is repulsed by Paul, yet she submits to his advances largely to avoid his aggression, eventually shifting from fear to pity. Here, the threats lie more in the potential dangers posed by men rather than the task of convincing them to have sex. If men were to play this mode of the game, it could add to the cringing realization of what their striving with women looked like from another perspective.

Romantic Encounters also included a feature called "Love Testing," which interviewed the player about hypothetical relationship dilemmas and offered insight into their companionate style. The game offered evaluations based on responses to diverse scenarios: What would the player do if he found his long-term heterosexual partner, to whom he has proposed marriage, has been having a long-running affair with another woman? How would he proceed if he were unsatisfied with the sexual aspects of an otherwise fulfilling new relationship? What would he do if an ex-girlfriend who is now married suggested that they start an affair? How does he react if his partner tells him she is leaving him because he is not ambitious enough?

The feature even diagnosed the user's issues with relationships. For example, depending on the player's responses to the "Love Challenge" test, they might be told:

> Based on your love capacity profile, we would encourage you to work on your self-esteem. Gradually begin to take risks at exposing your honest feelings. Responses to authentic disclosures run the gamut from anger and rejection to excitement and attraction. People who basically like themselves can not only tolerate any of these responses, but also have a greater potential for gratifying and committed relationships. A person who refused to respond from his real self hides behind a mask. In an extended relationship the mask slips, wears thin, eventually giving your partner confusing and disturbing glimpses of a man she never bargained on being with.

In diagnosing players, the love testing seems to focus on many of the issues with new masculinity the game interrogates, namely, how the player communicates with women and how he responds to challenges or situations he does not control. The love test seems to anticipate that some players in their real relationships would share the false, instrumental modes of relating to women that the player character shows in the game. Together, these different features show that *Romantic Encounters* was not only simulating heterosexual seduction as conquest narrative but also fostering reflection on shortcomings of the seemingly sensitive New Man.

Gaming Relationships

The relationship between players and avatars in video and computer games has received a lot of attention from game scholars for its role in shaping a style of masculinity that goes by different but related labels—gamer masculinity, technomasculinity, toxic geek masculinity.[69] When discussing player relationships to male avatars or player characters, the tendency is to discuss how these often hypermasculine figures function as an ego ideal. By playing the game through this avatar, male players are given access to experiences of power and mastery that they may feel they are otherwise denied outside the game. At times, this can apply directly to how male characters and players exert control or dominance over women. For example, in their discussion of masculine avatars in games, ranging from the hypermasculine Duke Nukem to geeky playboy Larry Laffer, Anastasia Salter and Bridget Blodgett argue "both suggest to gamers that the world is hostile to their desires, but through the right strategies or use of force their avatars can still 'score.'"[70]

Adult games have the potential to replicate these dynamics, reinforcing hegemonic gamer masculinity by simulating heterosexual relationships as playable scenarios. In fact, this appears to be the way members of the seduction community (i.e., pickup artists) see themselves in relation to heterosexuality and games. Ran Almog and Danny Kaplan analyzed seduction community discourses and found that they model pickup as "a game with predetermined rules that may be learned and practiced."[71] Moreover, Almog and Kaplan observe that in developing their skills in picking up women, men in the seduction community are encouraged to think of themselves as something like a game avatar. Like the player's relationship to a game avatar as ego ideal, pickup artists identify with and perform an idealized masculine persona to seduce women more effectively.

Yet, in video and computer games, the relationships between players and avatars or player characters are not always executed on an ego-ideal model. Some scholars have addressed games that rupture player identification or denaturalize masculinity. For example, Ewan Kirkland argues that the complex and contradictory way that masculinity is represented in the *Silent Hill* game series "deviates from the uncompromisingly macho, triumphantly aggressive, and uncritical narratives and expressions of masculinity with which video games are associated." Instead, it presents a "critical discourse on masculinity" that interrogates masculine domination, gendered violence, misogyny, heroic individualism, and agency in games.[72]

Ian Bryce Jones argues that fumblecore comedy games like *Octodad* and *QWOP* offer "a playful perversion of the expected relation between player and avatar" in which "games' otherwise monolithic trend toward corporeal power fantasy . . . begins to break down."[73] In fumblecore games, this occurs through control schemes that challenge the mapping of player control onto character actions and bring attention to the player's body. Discussing *Octodad* specifically, Bo Ruberg argues that the narrative and control schema in this game functions to parody "the ideal suburban, nuclear family and denaturalizes normative expectations around gender, sexuality, reproductivity, and the motions of the human body"—fostering a queer gaming experience by encouraging players to experience awkwardness.[74]

Existing scholarship has contributed to a better understanding of the dynamics by which games foster feelings of power or disempowerment through relationships to game avatars and even how this can foster a queer gaming experience. Still, there is more to parse in how heterosexuality is presented in games and how they situate players to make sense of romantic or sexual interactions. Mia Consalvo offers a take on this in her discussion of game romance where she uses Eve Sedgwick's formulation of erotic triangles of male homosocial desire to explore how presumed male players manage desire for game avatars.[75] She suggests a situation in which female love interests are used to divert potential desire that a presumed heterosexual male player may develop for the game avatar. In games, though, Consalvo suggests that the player is encouraged to collapse his identity with the male avatar, validating heterosexual masculinity more firmly.[76] This argument offers a perspective on how players are positioned in relation to heterosexuality, but it does not account for games in which the relationship between player and avatar is more fraught.

Focusing on the cringe aesthetics of adult games like *Leisure Suit Larry* and *Romantic Encounters* aids in reassessing how masculinity and heterosexuality function in computer and video games through relationships with masculine avatars or player characters. Specifically, the cringe aesthetics in these adult games critique masculinity by examining how a player character navigates social norms and relational modes rather than focusing on the body or player agency. Although *Leisure Suit Larry* and *Romantic Encounters* occasionally use physical abjection for humor, these games focus more on cringe effects by showing their player characters breaking social taboos in relationships with women characters and in performances of heterosexual seduction. They demonstrate how computers can inspire players to reflect on different modes of relationality. Unlike the pickup artists who see seduction as a game and model themselves on avatars as ego ideals of hegemonic masculinity, adult games have the potential to interrogate and critique this instrumental mode of relating to women.

CODA

Companionate Computing and Its Echoes

IN AN APPLE COMMERCIAL that aired on television in 1981, Apple spokesman Dick Cavett sits across from a woman working on an Apple II, whom he introduces as "an average American homemaker."[1] They are both at a desk in a sparsely decorated space—it is not clear if this is meant to represent a living room, home office, or any specific room of the house. Among the few decorations is a vase of flowers in the background, seemingly aiming for a feminine touch to counterbalance the large computer that serves as the center of attention. Repeating some of the uses of home computers often described in advertisements and the popular press of the time, Cavett asks this "average" woman if she is using her computer for household budgeting or recipe storage. To his surprise, she explains that in fact she is "working in gold futures" and can also use her Apple for "trend analyses" and the creation of bar graphs. Although Cavett is briefly taken aback, interrupting the woman to ask with astonishment, "Are you really a homemaker?," he quickly and humorously recovers, pronouncing to the viewer, "So . . . Apple is the appliance of the eighties for all those pesky household chores."

This commercial displays many of the conventional ways to frame computing in this period, working through some of the hopes and concerns about a technology that was relatively novel in households at this time. The implication that the Apple II did serious work like trend analyses and bar graphs aimed to distinguish this higher-priced computer from lower-priced competitors. But the ad also relied on and played with expectations about gendered computer use. It would be clear to viewers in 1981 that an "average American homemaker" would not own her own Apple II, since very few American homes had a computer at all. This woman engaging with onscreen bar graphs could even be interpreted as a response or correction to the first print ad for the Apple II, which could only imagine a woman in the background, doing meal prep while a man in the foreground generates line graphs on his new desktop. But here, the address of this commercial is confusing. Is the ad supposed to be appealing to women by including one in its vision of computing, only to call her average and suggest she may only transcend this averageness by using her computer to masculinize her work?

What interests me the most in this commercial, though, is how it jokingly contends with what it means for an average homemaker to use the computer in an appropriate way. The ultimate message seems to be that the computer is so powerful it can enhance home productivity by the standards of a masculine public world, rather than succumbing to any feminizing influence from women's domestic culture. In the end, the idea of computers used for household budgeting or recipe software by a truly "average" woman remains an absurd notion—at least in this commercial. In contrast, what this book has sought to argue is that this reverse relationship—in which computer culture is influenced by women's domestic culture and histories of negotiating companionate family relations in the home—is not absurd; in fact, it is an important aspect of the history of computing that has not been adequately addressed.

In this book, I have argued that some software and hardware makers, as they developed computing products for the home throughout the 1980s, were influenced by contemporary cultural changes that affected gender relations, domesticity, and the ideals of the companionate family— including changing conditions and norms that resulted from feminist social movements of the previous decades. As they developed computer products for the home, some in the industry conceived of ways to integrate applications of computing into domestic relationships, in some cases explicitly aiming to bolster companionate bonds rather than disrupt them. Through analysis of romance software that worked as a mediator of couples' relationships, robots that provided support and extension for a nurturing form of participant fatherhood, talking and listening dolls that served as proxies for parental and especially maternal care, and adult games that inspired critique of norms of heteromasculine relations with women, each chapter of this book uncovered a different mode through which computers augmented various companionate relationships. In other words, each chapter offered a different vision of what it would mean for computers to function as media of domestic relationality. In these ways, computer applications proposed to remediate companionate relationships as the target market of the white middle-class family experienced new pressures. Even so, as I have demonstrated, the attempts to use computing in this way were marked by tensions. These companionate relationships were burdened with many shifting expectations and responsibilities at a time of disorienting cultural and economic change. It was not always obvious how, or if, the computer would assuage these pressures—sometimes it even threatened to add burdens or alienate users from one another.

In these examples, I have tried to make palpable the impact of contemporaneous social forces, especially those related to women's changing roles in the home and public life, on computer products and cultures. The mechanism by which these cultural forces shaped computing varied, and the influence is not always straightforward or easy to detect. In the game *IntraCourse*, for example, contemporaneous changes in sexual culture made their way into the game partly through citation of previous decades of sex research. This program claimed a team of psychologists helped develop its advice and insights about relationships and sexuality, and it communicated a less restrictive view of the spectrum of sexual desires. On the other hand, one of the creators of *Lovers or Strangers* was influenced by women's movement and sexual revolution cultures in very different ways. Drawing on her years in sexual therapy as part of communal living, she designed the program to help women feel more comfortable discussing their own sexual pleasures with their partners. The influence of women's culture can also be seen in the marketing of this game, which mimicked romance novel aesthetics. In different ways, then, romance software was shaped by more general cultural contestations regarding sexuality and coupling, feminist and otherwise, even if these products did not all include women in their production.

Personal robots for the home seemed more firmly targeted at men and lacked women's involvement in their planning. Still, the robot firms, in their promotional discourses and discussions of home robots, presented these technologies as cute and companionable, comparing them to home appliances just as often as to computers. One engineer at Androbot even spoke of how the desire to make robots cute informed technological design decisions, although he suggested this was occasionally to their detriment.[2] Ideas about the home and men's place in it, then, influenced the ultimate design of these robots.

In the case of microprocessor-powered talking dolls, often hybrid ventures of established toy companies and electronics firms, their creation was informed at every level by histories of doll play and expectations for their anticipated role in girls' socialization. They were also shaped by shifting conceptions of mother–daughter relationships. Even a game like *Leisure Suit Larry*, seemingly emblematic of unchecked white masculine nerd culture, was developed in a context where game designers and marketers expressed concern about changing feminist standards for distinguishing sexist and nonsexist media. Software makers were also concerned with defining appropriate adult or pornographic representation suitable for the home environment. Their citations of the adult media industry, television, Hollywood, and Tipper Gore's censorship battles with the music industry are just a few indications of how contemporaneous discourses, both feminist and otherwise, informed the game's positioning.

By making these influences evident, *The Intimate Life of Computers* contributes to a larger project of feminist media histories and cultural histories of technology. It is not uncommon in media studies to account for the influence of domestic culture and companionate relations on the history of technologies like radio, television, phonography, and film.[3] This work is much less common in computer history, though some excellent scholarship has already explored the role of gender and sexuality. Jen Light has discussed how the work of early programmers or computer operators in the 1940s was aligned with feminized clerical labor. Nathan Ensmenger describes how, as the programming profession became more lucrative, programmers worked to masculinize its associations in the 1960s and 1970s. In that same period, Mar Hicks has discussed how gender and heteronormative assumptions played a role in defining and undercutting the labor practices of the British computer industry.[4]

But histories of computing as it developed in the late 1970s and 1980s as a personal medium for the home rarely do this. This may be because in that period not only were few women involved in the production of computer software and hardware, but the audience for computers was not often thought to include women in their address. These seem to be the conditions Laine Nooney is reacting to when she argues in her critique of the related field of video game history, "The common practice of 'adding women on' to game history in a gesture of inclusiveness fails to critically inquire into the ways gender is an infrastructure that profoundly affects who has access to what kinds of historical possibilities at a specific moment in time and space."[5] The approach in this book is in agreement with Nooney's argument that a shift in method is needed to include women as something more than "outliers." But whereas Nooney's work is "occupied with the historical mechanisms through which gendered bodies become legible to videogame history," my own approach focuses on how feminist critiques of companionate relations make their way into the history of computing even when it is difficult to find individual women involved in the production or consumption of this media.

Still, skeptical readers, even if they were convinced about the indirect ways feminist concerns influenced the products discussed in this book, may still potentially take issue with the choice of objects and their cultural or technical legacy. Admittedly, some of the examples in this book might seem quirky or quaint, and it is true they had varying degrees of commercial success at the time they were sold. For example, Galoob's Baby Talk was one of the best-selling toys in 1986, whereas Julie and Jill did not fare well. In fact, newspapers at the time claimed that Jill sold so poorly it nearly bankrupted Playmates. One company executive joked, "After Jill, the chairman won't even have an electronic doorbell in his house."[6] Similarly, Nolan Bushnell—who had become famous for founding Atari in the late 1970s but failed to find a market for his Androbot robots in the mid-1980s—later reported that he had lost $23 million of his own money on Androbot, which taught him "not to fall in love with your own product too much."[7] Reports suggest that *Interlude* sold relatively well at a time when the market for home computers was small. In contrast, *Lovers or Strangers* and *IntraCourse* do not seem to have succeeded commercially. In either case, these programs did not spark a lucrative field of romance software, although similar programs were released throughout the decade such as *The Marriage Counselor* (Human Perspectives, 1987) and *Treating Erection Problems* (Psycomp, 1984). Unlike many examples in this book, *Leisure Suit Larry* did well for its designer and publisher, inspiring many sequels and reboots.

Yet, although many of these products were not commercially successful, they represent a way of understanding computers as technologies of companionate relationality that has become widespread. This book began by discussing Amazon's Echo devices and Alexa virtual voice assistant, only one contemporary example of how companionate computing persists and has become naturalized. Digital media, in many different embodiments, are a familiar part of the contemporary middle-class home, involved in mediating even the most intimate aspects of family relationships.

While writing this book, many more echoes of 1980s companionate computing emerged or came to my attention. One could point to the compatibility features on dating apps as a potential parallel to the questionnaires about romance and sex found in romance software programs. Smart phones and other digital devices have become familiar tools for couples to negotiate their developing intimacies. Popular commentators have noted that sharing or synching digital calendars, playlists, streaming accounts, or passwords has become a milestone in coupled relationships.[8] A Pew Research study addressing this perception of digital media's importance in romantic relationships sought to itemize the different roles digital media play. It reported that 40 percent of coupled adults say they are at least sometimes bothered by the time their partner spends using their phone, half say their partner is sometimes distracted by the phone during a conversation, a third have looked through their partner's phone without their permission or knowledge, three-quarters share passwords, and a third use digital media to show affection or connect with their partner.[9] Even more proximate to examples of romance software from the 1980s are the various couples apps available for smartphones. Like romance software, apps like Coral or Paired function in a therapeutic way. They query existing couples about their desires and preferences and use this information to offer advice for how to increase sexual satisfaction or improve communication for couples who might be experiencing relationship challenges. This aligns with contemporaneous therapeutic culture, positioning the digital device as a mediator of these bonds.

Similarities extend beyond applications for romantic couples. In the 1980s, it was common for companies like Androbot and Hubot to promise that vacuuming robots were on the horizon. This promise has been realized in the millions of iRobot Roombas that populate U.S. homes. Chapter 2 documented the many instances in which Topo and Hero Jr. were depicted as cute pets and children by their designers and users. Studies of families using Roombas have painted a similar picture of how people relate to the robot. Some families name their Roombas, chat with them, and describe them as pets or "valuable family member[s]."[10]

The more ambitious plans for general purpose or programmable domestic robots are still being pursued. In fall 2021, Amazon unveiled its Astro home robot. If the commercials are to be believed, it fulfills many of the unmet promises of 1980s robot manufacturers.[11] It can follow users around, patrol their home for intruders, and even entertain children with dance (all features hearkening back to Heath's vision for the Hero Jr. in 1984). Astro is sold as cute and companionable, but unlike 1980s robots, it is not framed primarily as a tool men would use with their family. Astro users in Amazon's commercial comprise a diverse group, and the spot ends with the familiar image of a robot bringing its owner a beer—only this time she is a Black woman, not one of the endless parade of white men from 1980s robot ads.[12] Of course, as scholars like Thao Phan, Neda Atanasoski, and Kalindi Vora have argued, the actors in these promotions may have diversified, as have the owners of these robots, but these technologies still replicate a model of middle-class domesticity in which the robot stands in for racialized domestic service.[13] In this commercial as well, the narrative falls back on a conventional gendered pattern—the woman's eager husband tries to convince her through this beverage delivery that the robot has a use in their home.

Echoing talking dolls, Amazon commercials show how children's play can be enriched by the presence of their Echo Dot Kids. One such commercial features, among other representations, a young girl interested in space travel and shows how Alexa serves to foster her curiosity through educational play.[14] In this way, the promotional address for the Echo Dot Kids repeats some of the promises toy companies offered in the 1980s. This digital device serves as a proxy for parents, but in so doing incites similar anxieties about electronic babysitters. The ad concludes by reassuring parents that they ultimately have control over the device—Alexa can stand in for parental care, but it is no substitute. Of course, some aspects of this technology are quite different. Unlike the worries over Baby Talk, Julie, and Jill, many of the concerns about Echo Dot Kids relate to privacy and surveillance.[15] Furthermore, advertising for the device is less gender polarized.

Dating simulations are a more common genre of video and computer game today than they were in 1987 when *Leisure Suit Larry* and *Romantic Encounters* were released. Moreover, as scholars like Ran Almog and Danny Kaplan have discussed, communities of men have emerged that have literalized the connection between dating and video games by approaching relationships with women as challenges to manipulate and conquer using game logics.[16] At the intersection of dating sims and the seduction community is a fascinating, horrifying game created by the pickup artist Richard La Ruina called *Super Seducer* (2018), "the world's most realistic seduction simulator."[17] In the game, a sort of interactive branching video, the player is presented with different situations in which he might meet and pursue women: What do you do if you want to initiate an encounter with a beautiful woman on the street? How do you approach women at a bar? How do you turn a friend into a girlfriend? In each of the situations, players are given options to select from, and the game then plays a video clip to show how that selection would play out. The video plays whether the player has chosen what the game deems the right or wrong answer. Even the most obviously incorrect choices, involving immediate sexual propositioning or inappropriate touching, are acted out. Also, both right and wrong answers prompt direct explanation from La Ruina—flanked, for some unknown reason, by two half-naked women. It is difficult to tell if *Super Seducer* is meant at all earnestly as a training tool for pickup artists, as it claims. La Ruina is himself a leader in the seduction community, but the game is difficult to take seriously. Regardless of its intent, it replicates a dynamic of cringe and critique from *Leisure Suit Larry*. For those who play the game sincerely, the critique is limited to those scenes in which La Ruina performs his version of failed masculinity and gets rejected by women; for others, the game as a whole invites cringing reactions and reflections on masculine norms.

By noting echoes between the 1980s and more contemporary media, I am reframing the history of our present media environment, where pervasive digital media regularly intervenes in our companionate relationships.[18] In the 1980s, how we understand digital media's role as a companionate technology was first emerging, and many contemporary ways of using media have counterparts in that earlier period. Still, connecting these products may seem like an uneasy fit, for it may imply a claim of direct influence from the 1980s to today. Yet the practices or conceptions of computing described in this book do not appear to have inspired or influenced their contemporary counterparts—or at least that is not what this book has sought to trace.

So, if the objects analyzed in this book did not directly inspire their contemporary counterparts, what explains these echoes? Looking at romance software as just one illustration, the mode of interaction found in today's Coral app has many similarities to titles from the 1980s. In some ways, Coral could be viewed as a parallel to the sex-focused therapeutic intervention provided by *Interlude*. Coral was launched in November 2019 by Isharna Walsh, driven by her realization of her own dissatisfaction in a past sexual relationship. Coral claims that the app's methodology is overseen by a panel of experts, some of whom specialize in the sexual health and well-being of women, trans, and gender-nonconforming people, and those who have experienced sexual trauma. Like *Interlude*, Coral offers activities for couples that are intended to improve their relationships. For example, a Coral activity called "Foreplay all day!" directs users to spend a whole day seducing their partners: "It may seem indulgent to prioritize seduction, but that's the point!" This is quite like *Interlude*'s advice to play "Hooky Nookie," in which couples are told to take off work and devote an entire day to each other: "Now is your chance to say and do the things you always put off because you are tired or in a hurry."

In other guided sexual or physical activities in the Coral app, users follow audio directions with headphones while completing recommended techniques on their partners. These guided sound experiences are an example of how Coral tries to recruit the smartphone, which could distract from one's partner, into a coupled experience, similar to how *Interlude* attempted to mobilize desktop computers, which were also perceived as a potential distraction. The voice guidance in the Coral app is similar to the playful and sometimes kinky address the computer used to egg on users of *Interlude*. Of course, there are differences as well. *Interlude* was one of the romance programs that was especially heteronormative in its address, whereas Coral professes to be more expansive.[19] Furthermore, the logistics of using a smartphone as opposed to a desktop computer also play into the differences in how these technologies work.

Yet, despite some similarities, it is highly unlikely the makers of Coral had in mind titles like *Lovers or Strangers* or *Interlude* when they imagined how digital media could be used to bring couples together. Instead, the app seems to engage with contemporary anxieties about phones, understandings of sexual technique developed in therapeutic fields, ideas about sexuality that attempt to acknowledge nonbinary genders, and the stresses caused by couples navigating complex schedules. The similarities and differences between romance applications from the 1980s and today arise from their connection to ongoing discourses and traditions. Rather than suggest that these applications were inevitable or that specific technology from the 1980s directly influenced contemporary romance apps today, it seems instead that these echoes are the reverberations of tensions in companionate ideals that have shifted over time but remain unresolved. They resound with each new medium, but they originate from longer cultural threads and contestations over what a companionate family is and how it serves its members. It may not be particularly glamorous to focus on such mundane, repetitive technologies in the history of computing— a technology that on a grander scale is creating artificial intelligences, connecting people across oceans, and undermining democracies. Some might resist calling such technologies computing at all—someone once dismissively summarized my project to me as "Teddy Ruxpin studies."

Furthermore, what we find in these everyday technologies is not often something we would call feminist computing. Still, this approach helps underscore how the history of domestic relationality has long shaped the development of computer applications. Highlighting this influence is necessary to avoid writing gender, domesticity, and feminist culture out of the process of technological development and the history of computing.

ACKNOWLEDGMENTS

This project has developed over many years and benefited from the support and generosity of so many brilliant colleagues and friends. This research began when I was a graduate student in Screen Cultures at Northwestern University. I am grateful to my advisers Lynn Spigel and Jake Smith for their support and feedback. They have had a formative influence in shaping this work and my interest in media history. I would also like to thank Mimi White, Ariel Rogers, Jeff Sconce, Jackie Stewart, and Hamid Naficy for their mentorship and encouragement. In graduate school, I was so privileged to find friendships and intellectual community with other students who immediately made me feel like I had found my place. I feel so fortunate to have started my journey as an academic in your sparkling company. This includes Linde Murugan, Leigh Goldstein, Annie Sullivan, Catherine Harrington, Alla Gadassik, Hannah Spaulding, Alex Thimons, Simran Bhalla, Maureen Ryan, Ilana Emmett, Steve Babish, Chris Russel, Quinn Hartman, Zach Campbell, and Beatrice Choi. Annie Sullivan deserves an extra thank-you for her unending patience and incredible generosity as I developed my dissertation and in the years since.

Since arriving at Washington University in St. Louis, I have benefited from the support of my university colleagues, especially those in the Program in Film and Media Studies and Women, Gender, and Sexuality Studies. This includes Gaylyn Studlar, Rebecca Wanzo, Colin Burnett, Diane Wei-Lewis, John Powers, Jim Fleury, Chang-Min Yu, Raven Maragh-Lloyd, Ian Bogost, Julia Walker, Andrea Friedman, Heather Berg, and Mary Ann Dzuback. Colin Burnett has especially been an advocate and generous source of mentorship in the program. I am also grateful to have been provided a leave by the Program in Film and Media Studies that allowed me the much-needed time to research and write a large portion of this manuscript. I would also like to thank Pat Henry and Brett Smith.

I was lucky to spend a year after graduate school as a Mellon Postdoctoral Fellow at McGill. It was a privilege to see Jonathan Sterne in action as a supervisor and researcher and to take time to reformulate this project with the help of his insights. At McGill I would also like to thank Carrie Rentschler for generously listening and offering ideas and Alanna Thain for the opportunity to present research from this book for an Institute for Gender, Sexuality and Feminism workshop. I am especially grateful to Zoë De Luca for her friendship and intellectual engagement, often over oysters. She helped make Montreal a joyful experience.

So much of the research for this book was made possible by the incredible collections of computer magazines, emulations, and ephemera on the Internet Archive. Additionally, a grant from the Strong Museum of Play made it possible to consult invaluable resources relating to the toy, doll, and game industries. I would like to thank Julia Rossi, Patricia Hogan, Jon-Paul Dyson, Beth Lathrop, Tara Winner, and Christopher Bensch for making access to these collections possible. Near the end of this project, I had the rare opportunity to speak to Dr. Brigitta Olsen and Dr. Veronica Thomas, who were involved in the creation of some of the software discussed in chapters 1 and 4. I am so grateful that they took time to share their experiences with a stranger; I am honored to have heard their stories.

At so many points over the past few years I have been grateful to have colleagues who were generous enough to engage with this research in earlier stages and provide discerning feedback. An early draft of chapter 1 benefited from the generous comments of participants in the Women, Gender, and Sexuality Studies Colloquium; I am grateful to Rachel Brown and Rebecca Wanzo for organizing this opportunity and to Diane Wei-Lewis for responding. It was an honor to return to Northwestern and present material from chapter 1 at the outstanding Backwards Glances conference. I thank Cara Dickason for making this possible. Bo Ruberg generously read a draft of the entire manuscript and offered multiple hours of their time and attention to talk through ideas and strategize revisions. As I was completing the first draft, I was invigorated by the intellectual community provided by a writing group organized by Leigh Goldstein. In this group Linde Murugan, Melissa Phruksachart, Jaap Verheul, Brandy Monk-Payton, Curran Nault, Mimi White, and Moya Luckett provided invaluable feedback on multiple chapters but also served as inspiration by sharing their own brilliant work. At the University of Minnesota Press,

I would like to thank Leah Pennywark and Anne Carter for their support throughout the publication process.

I would like to offer a special note of gratitude to a few people without whom this project would not have ever been finished, or likely even started. As I was completing the book, Rebecca Wanzo was incredibly generous with her time, support, and feedback. She truly went beyond what I would ever expect from an academic mentor. Every chapter is better because of her insights and guidance. I hope I can continue to learn from her how to be a sharper thinker and supportive mentor. For more than a decade, Linde Murugan and Leigh Goldstein have been such a deep and essential source of support, insight, companionship, intellectual enrichment, and joy that it is impossible to imagine what this project (or my identity) would be without them. I am so grateful, and I hope one day to deserve their friendship. Lastly, I am sincerely grateful to my family—Nancy, Guis, Joe, and especially my mother—for sustaining me throughout this process and offering unconditional support in so many ways. I really could not have done any of this without them.

NOTES

Introduction

1. The World's Best Ads, "Amazon Alexa Super Bowl 2022 with Scarlett Johansson and Colin Jost," uploaded February 8, 2022, YouTube video, https://youtu.be/8bACuhV5RPM?.

2. Dave Lee, "Dave Limp: Amazon's Focus Is on the Real World, Not the Metaverse," *Financial Times,* April 22, 2022, https://www.ft.com/content/3d5bdfa8-d85e-41e8-b2d5-fadf8948cbba.

3. Sherry Turkle, *The Second Self: Computers and the Human Spirit* (New York: Simon and Schuster, 1984); Thomas Streeter, *The Net Effect: Romanticism, Capitalism, and the Internet* (New York: NYU Press, 2010); Elizabeth Petrick, "Imagining the Personal Computer: Conceptualizations of the Homebrew Computer Club 1975–1977," *IEEE Annals of the History of Computing* 39, no. 4 (October–December 2017): 27–39; Luke Stark, "Here Come the 'Computer People': Anthropomorphosis, Command, and Control in Early Personal Computing," *IEEE Annals of the History of Computing* 42, no. 4 (October–December 2020): 53–70.

4. See for example Lynn Spigel, *Make Room for TV: Television and the Family Ideal in Postwar America* (Chicago: University of Chicago Press, 1992); Michele Hilmes, *Radio Voices: American Broadcasting, 1922–1952* (Minneapolis: University of Minnesota Press, 1997); Keir Keightley, "'Turn It Down!' She Shrieked: Gender, Domestic Space, and High Fidelity, 1948–59," *Popular Music* 15, no. 2 (May 1996): 149–77; Carolyn Marvin, *When Old Technologies Were New: Thinking about Electric Communication in the Late Nineteenth Century* (New York: Oxford University Press, 1988); Lisa Gitelman, *Always Already New: Media, History, and the Data of Culture* (Cambridge, Mass.: MIT Press, 2006); Jonathan Sterne, *The Audible Past: Cultural Origins of Sound Reproduction* (Durham, N.C.: Duke University Press, 2003).

5. See, for example, Kathryn Harris, "Home Computer Prices Fall—but Where's the Market?," *Los Angeles Times*, January 13, 1981, sec. 4; David E. Sanger, "The Expected Boom in Home Computers Fails to Materialize," *New York Times*, June 4, 1984. For more on the home computer as product category, see Michael L. Black, *Transparent Designs: Personal Computing and the Politics of User-Friendliness* (Baltimore: Johns Hopkins University Press, 2022); Jimmy Maher, "Business Is War," *Digital Antiquarian* (blog), December 20, 2012, https://www.filfre.net/2012/12/business-is-war/.

6. David E. Sanger, "Home Computer Is Out in the Cold," *New York Times*, February 20, 1985.

7. Dan Rapoport, "Christmas Future: The Computers Are Coming! Or Are They?," *Washington Post*, December 4, 1977, sec. Magazine; Matt Seiden, "Home Computer Revolution Is on Hold," *The Sun*, July 3, 1985.

8. Although potentially read as an offhand joke for their Christmas issue, the *Byte* editors later said this was the first time they received reader feedback on a cover image. "In This Byte," *Byte*, December 1975.

9. See, for example, Stephanie Mansfield, "A Computer Is the Apple of His Eye: Byte-ing into Marital Bliss," *Washington Post*, November 22, 1982; Art Buchwald, "Lonely Nights for the Poor Computer Widow," *Boston Globe*, June 29, 1982; Joann S. Lublin, "Marriage a La Modem: A Computer Widow's Lament," *Wall Street Journal*, August 12, 1983; Sarah Kortum, "How I Learned about Computers to Save My Marriage! The Story of a Reformed Computer Widow," *Family Computing*, April 1984; Lyn Chase, "Computer Widow's Compendium," *Guide to Computer Living*, May 1986; J. D. Sidley, "Light Touch: Paradise Lost," *Family Computing*, November 1984. Stories of husbands losing their wives to computing were much more rare, but for an example of this see Kathy Chin, "Female Hackers Turn Husbands into Microcomputer Widowers," *InfoWorld*, April 11, 1983. For further discussion of discourses about computer widows, see Lori Reed, "Domesticating the Personal Computer: The Mainstreaming of a New Technology and the Cultural Management of a Widespread Technophobia, 1964–," *Critical Studies in Media Communication* 17, no. 2 (June 2000): 159–85.

10. Georgia Dullea, "New Marital Stress: The Computer Complex," *New York Times*, January 10, 1983; John Markoff, "Computing in America: A Masculine Mystique," *New York Times*, February 13, 1989; Jean Dietz, "Logged On, Turned Off: Some Are Afraid of Computers, Some Are Obsessed with Them," *Boston Globe*, February 6, 1984; Turkle, *Second Self.*

11. Reed, "Domesticating the Personal Computer," 177.

12. Steven Levy, *Hackers: Heroes of the Computer Revolution*, 25th anniv. ed. (Sebastopol, Calif.: O'Reilly Media, 2010); Michael Swaine and Paul Freiberger, *Fire in the Valley: The Birth and Death of the Personal Computer*, 3rd ed. (Dallas: Pragmatic Bookshelf, 2014).

13. Black, *Transparent Designs*. See also Jesse Adams Stein, "Domesticity, Gender and the 1977 Apple II Personal Computer," *Design and Culture* 3, no. 2 (2011): 193–216.
14. Lori Emerson, *Reading Writing Interfaces: From the Digital to the Bookbound* (Minneapolis: University of Minnesota Press, 2014), xi–xii.
15. Laine Nooney, *The Apple II Age: How the Computer Became Personal* (Chicago: University of Chicago Press, 2023), 191.
16. Martin Campbell-Kelly, William Aspray, Nathan Ensmenger, and Jeffrey R. Yost, *Computer: A History of the Information Machine*, 3rd ed. (Boulder, Colo.: Westview Press, 2014), 235.
17. James Hay, "Unaided Virtues: The (Neo-)Liberalization of the Domestic Sphere," *Television and New Media* 1, no. 1 (February 2000): 53–73.
18. Campbell-Kelly et al., *Computer*, chapter 11.
19. Fred Turner, *From Counterculture to Cyberculture: Stewart Brand, the Whole Earth Network, and the Rise of Digital Utopianism* (Chicago: University of Chicago Press, 2006); Streeter, *Net Effect*; Thierry Bardini, *Bootstrapping: Douglas Engelbart, Coevolution, and the Origins of Personal Computing* (Stanford, Calif.: Stanford University Press, 2000).
20. Streeter, *Net Effect*, 17–43.
21. Streeter, 71, 88.
22. Streeter, 79.
23. Joy Lisi Rankin, *A People's History of Computing in the United States* (Cambridge, Mass.: Harvard University Press, 2018).
24. Robert Kominski, *Current Population Reports, Series P-23, No. 155: Computer Use in the United States, 1984* (Washington, D.C.: U.S. Bureau of the Census, 1988).
25. Justine Cassell and Henry Jenkins, "Chess for Girls? Feminism and Computer Games," in *From Barbie to Mortal Kombat: Gender and Computer Games*, ed. Justine Cassell and Henry Jenkins (Cambridge, Mass.: MIT Press, 1998), 2–45; Janet Abbate, *Recoding Gender: Women's Changing Participation in Computing* (Cambridge, Mass.: MIT Press, 2012), 2.
26. Kominski, *Current Population Reports*.
27. Dieter Bohn, "Exclusive: Amazon Says 100 Million Alexa Devices Have Been Sold—What's Next?," *The Verge*, January 4, 2019, https://www.theverge.com/2019/1/4/18168565/amazon-alexa-devices-how-many-sold-number-100-million-dave-limp.
28. "History—iRobot," accessed June 14, 2022, https://www.irobot.com.au/About-iRobot/Company-information/History.
29. An extraordinary exception to this is the ECHO IV (Electronic Computing Home Operator) computer, custom built by James Sutherland for his own Pittsburg home in 1966. The ECHO IV weighed eight hundred pounds and consisted of four cabinets measuring six feet by six feet by two feet, constructed in the family's basement. Sutherland built the ECHO IV with spare parts from

his employer, Westinghouse; it was used as a public relations tool for the company. For discussion of the ECHO IV, see James E. Tomayko, "Electronic Computer for Home Operation (ECHO): The First Home Computer," *IEEE Annals of the History of Computing* 16, no. 3 (Autumn/Fall 1994): 59–61; Dag Spicer, "If You Can't Stand the Coding, Stay Out of the Kitchen: Three Chapters in the History of Home Automation," *Dr. Dobb's, The World of Software Development* (blog), August 12, 2000, http://www.drdobbs.com/architecture-and-de sign/if-you-cant-stand-the-coding-stay-out-of/184404040; Dag Spicer, "The ECHO IV Home Computer: 50 Years Later," *CHM Blog*, May 31, 2016, https:// computerhistory.org/blog/the-echo-iv-home-computer-50-years-later/.

30. Paul E. Ceruzzi, *A History of Modern Computing*, 2nd ed. (Cambridge, Mass.: MIT Press, 2003), 13–78.

31. Ceruzzi, 109–42.

32. Paul Ceruzzi, "From Scientific Instrument to Everyday Appliance: The Emergence of Personal Computers, 1970–77," *History and Technology* 13 (1996): 3. This article provides a thoughtful discussion of the relative influence of technological development and social factors in the development of personal computing.

33. Turner, *From Counterculture to Cyberculture*, 104.

34. Rankin, *People's History of Computing in the United States*.

35. Turner, *From Counterculture to Cyberculture*, 2; Gitelman, *Always Already New*, 92.

36. Ceruzzi, "From Scientific Instrument to Everyday Appliance." For more on the history of microprocessors, see Michael S. Malone, *The Microprocessor: A Biography* (Santa Clara, Calif: TELOS/Springer-Verlag, 1995).

37. The Altair was not the first microprocessor-based kit computer offered commercially, but it is considered the first to gain influence with hobbyist users. See Ceruzzi, *History of Modern Computing*, 224–26.

38. H. Edward Roberts and William Yates, "Exclusive! Altair 8800: The Most Powerful Minicomputer Project Ever Presented—Can Be Built for under $400," *Popular Electronics*, January 1975, 33, 38.

39. Instead, the user had to program the device using small toggles on the front panel and interpret the computer's state through blinking lights. Ceruzzi, "From Scientific Instrument to Everyday Appliance." For more on the Homebrew Computer Club and their approach to personal computing, see Petrick, "Imagining the Personal Computer."

40. For more on design and advertising of the Apple II as a personal appliance, see Stein, "Domesticity, Gender and the 1977 Apple II Personal Computer."

41. Lawrence C. Levy, "Soon, the Home Computer," *Boston Globe*, April 9, 1978.

42. Myron Berger, "The Home Computer Rapidly Is Becoming a Household Word," *Chicago Tribune*, December 6, 1981; "Home Computers Come of Age," *New*

York Times, April 25, 1982; Don Nunes, "Once the Games Are Played Out, the Real Fun Begins," *Hartford Courant,* November 24, 1982.

43. Alexander Auerbach, "Home Computer Boom Is a Laugh: Practical Use Small but Fun Lures Buyers," *Los Angeles Times,* June 11, 1978.

44. Peter J. Schuyten, "Home Computer: Demand Lags," *New York Times,* June 7, 1979.

45. Ceruzzi, *History of Modern Computing,* 268–69.

46. For a history of word processing software, see Thomas J. Bergin, "The Origins of Word Processing Software for Personal Computers: 1976–1985," *IEEE Annals of the History of Computing* 28, no. 4 (October–December 2006): 32–47. For a discussion of the history of word processing in relation to literary authorship, see Matthew G. Kirschenbaum, *Track Changes: A Literary History of Word Processing* (Cambridge, Mass.: Harvard University Press, 2016).

47. Kathryn Harris, "Home Computer Prices Fall—but Where's the Market?," *Los Angeles Times,* January 13, 1981.

48. Robert X. Cringely as quoted in Ted Friedman, *Electric Dreams: Computers in American Culture* (New York: NYU Press, 2005), 104.

49. Steven Ditlea, "When a Computer Joins the Family," *New York Times,* August 30, 1979, sec. Home; Sanger, "Expected Boom in Home Computers Fails to Materialize."

50. Starting in Christmas 1982, prices for low-end computers plummeted: the TI-99/4A was priced at $200 after a rebate and the Commodore VIC-20 at $175. By fall of 1983, users could purchase the more powerful Commodore 64 for $200, the VIC-20 for only $80 and eventually, after TI announced it was abandoning the home computer market, the TI-99/4A for only $50. See Maher, "Business Is War."

51. Kominski, *Current Population Reports,* 1–2.

52. Michael deCourcy Hinds, "Get Your Program, Folks: It's a Chipper New World in Home Computers," *Chicago Tribune,* June 21, 1981, sec. N.

53. Alladi Venkatesh, Eric Shih, and Norman Stolzoff, "A Longitudinal Analysis of Computing in the Home: Based on Census Data, 1984–1997," in *Home Informatics and Telematics: Information, Technology and Society,* ed. Andy Sloane and Felix van Rijn (Berlin: Springer Publishing, 2000), 205–15.

54. For an example of the use of the phrase "computer in every closet," see Christine Winter, "Firms Get the Signal: Personal Computers Are Right at Home," *Chicago Tribune,* February 3, 1983, sec. 4.

55. Abbate, *Recoding Gender,* 2; Cassell and Jenkins, "Chess for Girls?," 10–14.

56. Reed, "Domesticating the Personal Computer."

57. Buchwald, "Lonely Nights for the Poor Computer Widow"; Patricia Walsh, "Tempo: Computer Widow: Her Marriage Is Off-Line," *Chicago Tribune,* January 12, 1984, sec. NW.

58. Dullea, "New Marital Stress." See also Michael Ferrari et al., "Home Computers: Implications for Children and Families," *Marriage and Family Review* 8, nos. 1–2 (June 5, 1985): 41–57.

59. The magazine changed its name to *Family and Home Office Computing* beginning with the October 1987 issue.

60. Karen Klassen, "Castles, Cathedrals, and Computers," *Family Computing*, November 1983; Deb Di Gregorio, "Celebrate! Let Your Computer Turn Party Planning into a Piece of Cake," *Family Computing*, July 1984; Jeanne Choffee, "Making Up Your Mind with *VisiCalc*," *Family Computing*, November 1983.

61. Robin Raskin, "How to Throw the Best Birthday Party on Your Block!," *Family Computing*, June 1986; Ralph Blois, "Family Computing: What's in a Name? A Business! How to Create a Family Newsletter for Fun (and Profit)," *Family and Home Office Computing*, March 1988.

62. "Announcing Family Computing's 1st 'Computing Family of the Year' Contest," *Family Computing*, November 1984, 71.

63. Linda Williams, "1985 Computing Family of the Year," *Family Computing*, September 1985, 42.

64. Steven Mintz and Susan Kellogg, *Domestic Revolutions: A Social History of American Family Life* (New York: Free Press, 1988), 107–31.

65. Natasha Zaretsky, *No Direction Home: The American Family and the Fear of National Decline, 1968–1980* (Chapel Hill: University of North Carolina Press, 2007), 2–17.

66. Alice Leppert, *TV Family Values: Gender, Domestic Labor, and 1980s Sitcoms* (New Brunswick, N.J.: Rutgers University Press, 2019), 1–2.

67. Arlie Hochschild, *The Second Shift: Working Parents and the Revolution at Home* (New York: Viking, 1989), 2; Zaretsky, *No Direction Home*, 241.

68. Hochschild, *Second Shift*; Zaretsky, *No Direction Home*, 10–11; Kay Longcope, "New Notions about Parenting," *Boston Globe*, September 28, 1982.

69. Alice Leppert, "Solving the Day-Care Crisis, One Episode at a Time: Family Sitcoms and Privatized Child Care in the 1980s," *Cinema Journal* 56, no. 2 (2017): 67–72. In addition to discussing the conditions of the daycare crisis, Leppert analyzes how sitcoms in the 1980s responded to these cultural anxieties.

70. Hochschild, *Second Shift*, 12.

71. Susan Faludi, *Backlash: The Undeclared War against American Women* (New York: Crown, 1991).

72. See, for example, Jane Marks, "The Supermom Myth, or Why Nobody's Perfect," *Parents*, February 1980; Lynn Langway, "Bringing Up Superbaby," *Newsweek*, March 28, 1983.

73. E. Anthony Rotundo, "Patriarchs and Participants: A Historical Perspective on Fatherhood in the United States," in *Beyond Patriarchy: Essays by Men on*

Pleasure, Power, and Change, ed. Michael Kaufman (New York: Oxford University Press, 1987), 74.

74. Kirsten Swinth, *Feminism's Forgotten Fight: The Unfinished Struggle for Work and Family* (Cambridge, Mass.: Harvard University Press, 2018).

75. David R. Shumway, *Modern Love: Romance, Intimacy, and the Marriage Crisis* (New York: NYU Press, 2003).

76. Jane Juffer, *At Home with Pornography: Women, Sexuality, and Everyday Life* (New York: NYU Press, 1998).

77. Hay, "Unaided Virtues," 63.

78. Shumway, *Modern Love*; Kristin Celello, *Making Marriage Work: A History of Marriage and Divorce in the Twentieth-Century United States* (Chapel Hill: University of North Carolina Press, 2009).

79. Elizabeth A. Patton, *Easy Living: The Rise of the Home Office* (New Brunswick, N.J.: Rutgers University Press, 2020), chapters 12–13.

1. A Ménage à Trois with Your Computer

1. Ted Friedman, *Electric Dreams: Computers in American Culture* (New York: NYU Press, 2005); Thomas Streeter, *The Net Effect: Romanticism, Capitalism, and the Internet* (New York: NYU Press, 2010); Luke Stark, "Here Come the 'Computer People': Anthropomorphosis, Command, and Control in Early Personal Computing," *IEEE Annals of the History of Computing* 42, no. 4 (October–December 2020): 53–70.

2. See, for example, Sylvia Scott, "My Husband's Computer Was My Competition," *Ladies' Home Journal,* January 1982; Art Buchwald, "Lonely Nights for the Poor Computer Widow," *Boston Globe,* June 29, 1982; Philip Faflick, "The Real Apple of His Eye: How Families Come Apart in the Face of the Micro Invasion," *Time,* August 30, 1982; Joann S. Lublin, "Marriage a La Modem: A Computer Widow's Lament," *Wall Street Journal,* August 12, 1983; Lyn Chase, "Computer Widow's Compendium," *Guide to Computer Living,* May 1986; J. D. Sidley, "Light Touch: Paradise Lost," *Family Computing,* November 1984. An extended discussion of the figure of the computer widow can be found in Lori Reed, "Domesticating the Personal Computer: The Mainstreaming of a New Technology and the Cultural Management of a Widespread Technophobia, 1964–," *Critical Studies in Media Communication* 17, no. 2 (June 2000): 159–85.

3. Sidley, "Light Touch," 42.

4. John D'Emilio and Estelle B. Freedman, *Intimate Matters: A History of Sexuality in America,* 3rd ed. (Chicago: University of Chicago Press, 2012), chapter 14; Barbara Ehrenreich, Elizabeth Hess, and Gloria Jacobs, *Re-making Love: The Feminization of Sex* (Garden City, N.Y.: Anchor Press / Doubleday, 1986); Steven Mintz and Susan Kellogg, *Domestic Revolutions: A Social History of American Family Life* (New York: Free Press, 1988), 203–38.

5. Arlie Hochschild, *The Second Shift: Working Parents and the Revolution at Home* (New York: Viking, 1989); Kirsten Swinth, *Feminism's Forgotten Fight: The Unfinished Struggle for Work and Family* (Cambridge, Mass.: Harvard University Press, 2018).

6. Jane F. Gerhard, *Desiring Revolution: Second-Wave Feminism and the Rewriting of American Sexual Thought, 1920 to 1982* (New York: Columbia University Press, 2001).

7. Elana Levine, *Wallowing in Sex: The New Sexual Culture of 1970s American Television* (Durham, N.C.: Duke University Press, 2007), 1–2.

8. Jane Juffer, *At Home with Pornography: Women, Sexuality, and Everyday Life* (New York: NYU Press, 1998), chapter 2.

9. Ehrenreich, Hess, and Jacobs, *Re-making Love*, 81.

10. Ehrenreich, Hess, and Jacobs, 104.

11. David R. Shumway, *Modern Love: Romance, Intimacy, and the Marriage Crisis* (New York: NYU Press, 2003), 27. See also chapter 5.

12. Shumway, *Modern Love;* Kristin Celello, *Making Marriage Work: A History of Marriage and Divorce in the Twentieth-Century United States* (Chapel Hill: University of North Carolina Press, 2009); Hochschild, *Second Shift;* Mimi White, *Tele-advising: Therapeutic Discourse in American Television* (Chapel Hill: University of North Carolina Press, 1992); Anthony Giddens, *The Transformation of Intimacy: Sexuality, Love and Eroticism in Modern Societies* (Stanford, Calif.: Stanford University Press, 1992).

13. Hochschild, *Second Shift.*

14. Celello, *Making Marriage Work*, 143–44.

15. Giddens, *Transformation of Intimacy*, 58.

16. Mintz and Kellogg, *Domestic Revolutions*, 204.

17. Sarah Kortum, "How I Learned about Computers to Save My Marriage!: The Story of a Reformed Computer Widow," *Family Computing*, April 1984, 38.

18. Kortum, 38–40.

19. Bebe Moore Campbell, "Computers and You: 'How My Computer Has Put *Punch* in My Marriage!'," *Essence*, July 1986, 98.

20. Mindy Pantiel and Becky Petersen, "A Happy Marriage: A Husband–Wife Team Combines Her Craft with His Computer Skills," *Family Computing*, November 1984; Diane Walkowiak, "*Softalk*, the Matchmaker, Letter to the Editor," *Softalk*, November 1981.

21. Bob Greene, "Computer Shows How to Be a Real Operator," *Chicago Tribune*, November 5, 1980, B1.

22. Syntonic, "*Interlude*, How's Your Love Life?," advertisement, *InfoWorld*, May 11, 1981, 17.

23. Robert Kominski, *Current Population Reports, Series P-23, No. 155: Computer Use in the United States, 1984* (Washington, D.C.: U.S. Bureau of the Census, 1988); Christine Winter, "Be It Ever So Humble . . . Computer in the Home?," *Chicago Tribune*, January 31, 1979.

24. Greene, "Computer Shows How to Be a Real Operator"; Chris Brown, "Review of *Interlude*: The Ultimate Experience, Syntonic Software Corp.," *80 Microcomputing*, September 1980, 21–22; John Barry, "Sexually Explicit Software: Computers in the Bedroom," *InfoWorld*, October 12, 1981, 30.

25. Philip Tubb, "*Interlude* from Syntonic Software Corp.," *InfoWorld*, October 13, 1980, 13.

26. See, for example, Luke Plunkett, "Sex Programs Have Come a Long Way since 1986," *Kotaku*, January 2, 2017, https://kotaku.com/sex-programs-have-come -a-long-way-since-1986-1790675408; and Jimmy Maher, "Sex Comes to the Micros," *Digital Antiquarian* (blog), February 27, 2012, https://www.filfre .net/2012/02/sex-comes-to-the-micros/.

27. Eric Grevstad, "E.G. For Example: Erotic Software," *inCider*, November 1986, 150.

28. Syntonic Software, *Interlude Manual*, 1980, https://archive.org/details/Inter lude_1980_Syntonic_Software_Corp_a, vii.

29. Unless otherwise noted, my analysis of *Interlude* is based on a browser-based emulation of the program on the Internet Archive—available at https://ar chive.org/details/Interlude_i—and the previously cited scanned instruction manual, Syntonic Software, *Interlude Manual*.

30. Syntonic Software, *Interlude Manual*, viii.

31. Ehrenreich, Hess, and Jacobs, *Re-making Love*.

32. Syntonic Software, *Interlude Manual*, viii.

33. Brown, "Review of *Interlude*," 22.

34. Greene, "Computer Shows How to Be a Real Operator," B1.

35. Barry, "Sexually Explicit Software," 30.

36. Another ad for *Interlude II* showed a woman on the phone with a visually absent partner but still focused on the dialogue between them.

37. Mary A. C. Fallon, "With This Computer Software I Thee Wed," *Des Moines Register*, May 17, 1984.

38. Brigitta Olsen, email message to author, August 21, 2023.

39. Olsen.

40. Fallon, "With This Computer Software I Thee Wed."

41. Janice A. Radway, *Reading the Romance: Women, Patriarchy, and Popular Literature* (1984; repr., Chapel Hill: University of North Carolina Press, 1991), chapter 3.

42. Screenshots from an emulation of *Lovers or Strangers* on the MobyGames online software collection—available at https://www.mobygames.com/game/ lovers-or-strangers—suggest that users are asked to provide their own choice before anticipating their partner's answer. The emulation of the reissue called *Friends or Lovers* on the Internet Archive—available at https://archive.org/ details/a2woz_Friends_or_Lovers—shows the reverse order.

43. White, *Tele-advising*, 53, 63.

44. "*Lovers or Strangers:* Creative Computing Software Profile," *Creative Computing Software Buyer's Guide 1983,* 1983, 69, https://archive.org/details/Creative ComputingBuyersGuide1983.

45. Olsen, email message to author.

46. Intracorp, *IntraCourse Manual,* 1986, II, author's collection. Note that the mix of roman numerals and letters used as page markers in the citations for this source reflect the original manual's informal style.

47. Intracorp, "*IntraCourse:* High Tech Sex," advertisement, *Ahoy!,* December 1986, 9, https://archive.org/details/ahoy-magazine-36/page/n7/mode/2up. For more on public fascination with the Kinsey reports, see Sarah E. Igo, *The Averaged American: Surveys, Citizens, and the Making of a Mass Public* (Cambridge, Mass.: Harvard University Press, 2009), chapters 5–6.

48. Intracorp, *IntraCourse Manual,* I.

49. Intracorp, f.

50. Leigh Mathia Goldstein, "TV for Women Who Think: Female Intellectualism and Network Television in Mid-Century America" (PhD diss., Northwestern University, 2020), 74–83.

51. Intracorp, *IntraCourse Manual,* III.

52. Intracorp, I.

53. Password protection developed as early as the 1960s to manage access to time-sharing systems at MIT and Bell Labs. See Brian Lennon, "The Long History, and Short Future, of the Password," *The Conversation,* May 3, 2017, http://theconversation.com/the-long-history-and-short-future-of-the-password-76690; Robert Morris and Ken Thompson, "Password Security: A Case History," *Communications of the ACM* 22, no. 11 (1979): 594–97; and Robert McMillan, "The World's First Computer Password? It Was Useless Too," *Wired,* January 27, 2012, https://www.wired.com/2012/01/computer-password/.

54. Richard Herring, "Review of *IntraCourse,* Intracorp, Inc.," *Ahoy!,* October 1986.

55. Syntonic Software, *Interlude Manual,* vii.

56. Syntonic Software, *Interlude* advertisement, *Creative Computing,* August 1980, 89.

57. Fallon, "With This Computer Software I Thee Wed," 39.

58. Fallon.

59. Intracorp, "*IntraCourse.*"

60. *IntraCourse* Sharper Image insert advertisement, *Popular Science,* April 1986.

61. Rebecca L. Davis, *More Perfect Unions: The American Search for Marital Bliss* (Cambridge, Mass.: Harvard University Press, 2010), 26, 102.

62. Giddens, *Transformation of Intimacy,* 58.

63. Bill Howard, "Abort, Retry, Ignore," *PC Magazine,* January 13, 1987, 152.

64. "*Lovers or Strangers,*" 66, 69.

65. "Reviews of New Products: *Lovers or Strangers,* Computer Game (for Apple II), Alpine Software," *Arcade Express,* December 19, 1982.

66. *IntraCourse* Sharper Image insert advertisement.
67. Herring, "Review of *IntraCourse*," 68.
68. Mar Hicks, "Computer Love: Replicating Social Order through Early Computer Dating Systems," *Ada: A Journal of Gender, New Media, and Technology*, no. 10 (2016): 6.
69. Donna J. Drucker, "Keying Desire: Alfred Kinsey's Use of Punched-Card Machines for Sex Research," *Journal of the History of Sexuality* 22, no. 1 (January 2013): 107.

2. "Not an Appliance, but a Friend"

1. An example of this advertisement can be found in Heath, "Take Your Family beyond Computers," *Modern Electronics*, December 1984, 29.
2. See, for example, Robert L. Simison, "R2D2 May Be a Wiz, but Robots for Home Use Still Can't Do Much," *Wall Street Journal*, April 25, 1985; and Robin Webster, "A Personal Robot Goes to Market," *PC Magazine*, August 21, 1984, 141.
3. See, for example, "Introducing the New Topo—Advertisement," *InfoWorld*, December 26, 1983/January 2, 1984, 26–27.
4. Darla Miller, "R2D2, We Want *You* in the Home," *Boston Globe*, May 24, 1980, 9; Cynthia Gorney, "Robotic Pets: Not Yet R2D2: And Please Don't Ask If They Do Windows," *Washington Post*, September 9, 1985; Michael Alexander, "Personal Robots Still a Long Way from R2D2," *Boston Globe*, April 11, 1986; Robert L. Simison, "Home Electronics: Personal Robots Don't Fulfill Original Vision," *Chicago Tribune*, May 5, 1985, sec. 15.
5. Tom Belden, "A Friendly Little Robot for the House," *Boston Globe*, December 27, 1982, sec. Business; "Inviting the Robot for Cocktails," *New York Times*, January 16, 1983.
6. Elizabeth A. Patton, *Easy Living: The Rise of the Home Office* (New Brunswick, N.J.: Rutgers University Press, 2020), 152.
7. Patton, 153.
8. An example of this advertisement can be found in *Byte*, July 1977, 22–23, https://archive.org/details/byte-magazine-1977-07.
9. Marsha F. Cassidy, "Cyberspace Meets Domestic Space: Personal Computers, Women's Work, and the Gendered Territories of the Family Home," *Critical Studies in Media Communication* 18, no. 1 (March 2001): 44–65; Patton, *Easy Living*.
10. James A. Levine, "A Computer in the Family," *Parents*, July 1983.
11. Levine, 57.
12. Kirsten Swinth, *Feminism's Forgotten Fight: The Unfinished Struggle for Work and Family* (Cambridge, Mass.: Harvard University Press, 2018), chapter 2.
13. Barbara Ehrenreich, *Fear of Falling: The Inner Life of the Middle Class* (New York: Harper Perennial, 1989), 206–7.

14. Robert Ayres and Steve Miller, "The Impacts of Industrial Robots" (Robotics Institute, Carnegie Mellon University, November 1981); *Robotics: Hearings before the Subcommittee on Investigations and Oversight of the Committee on Science and Technology U.S. House of Representatives*, 97th Cong. (1982) (statement of Paul H. Aron, Daiwa Securities America Inc.).

15. See, for example, Otto Friedrich, Christopher Redman, and Janice C. Simpson, "The Robot Revolution: For Good or Ill, It Is Already Transforming the Way the World Works," *Time*, December 8, 1980; Tom Nicholson, Kim Willenson, and Ayako Doi, "Robots: Japan Takes the Lead," *Newsweek*, September 21, 1981; Harry Anderson et al., "Where the Jobs Are—and Aren't," *Newsweek*, November 23, 1981; and William D. Marbach, Hope Lampert, and William J. Cook, "The Factory of the Future," *Newsweek*, September 6, 1982.

16. Amy Sue Bix, *Inventing Ourselves Out of Jobs? America's Debate over Technological Unemployment, 1929–1981* (Baltimore: Johns Hopkins University Press, 2000), 281.

17. Ehrenreich, *Fear of Falling*, 205–10.

18. Arlie Hochschild, *The Second Shift: Working Parents and the Revolution at Home* (New York: Viking, 1989).

19. Swinth, *Feminism's Forgotten Fight*, 42–69; Robert Charm, "From Macho to Mellow: Conference on Masculinity Explores Liberation, Male Roles," *Boston Globe*, July 5, 1981.

20. E. Anthony Rotundo, "Patriarchs and Participants: A Historical Perspective on Fatherhood in the United States," in *Beyond Patriarchy: Essays by Men on Pleasure, Power, and Change*, ed. Michael Kaufman (New York: Oxford University Press, 1987), 74.

21. Swinth, *Feminism's Forgotten Fight*, 42–69. Of course, not all men's liberationists were pro-feminist, and this period was also marked by a burgeoning men's rights movement in reaction to feminist gains of the 1970s.

22. Rotundo, "Patriarchs and Participants," 76–77.

23. Nadine Brozan, "New Marriage Roles Make Men Ambivalent about Fatherhood," *New York Times*, May 30, 1980; Ellen Goodman, "At Large: Fatherhood Reinvented," *Hartford Courant*, June 10, 1980.

24. Rotundo, "Patriarchs and Participants," 76–78.

25. Jon Patrick Harper, "Life with *Bachelor* Father and His Computer," *Family Computing*, November 1984, 51–56.

26. Kathy Chin, "Home Is Where the Job Is: Is Telecommuting the Work Style of the Future or Just Another Fad?," *InfoWorld*, April 23, 1984, 35.

27. Andrew Pollack, "Rising Trend of Computer Age: Employees Who Work at Home," *New York Times*, March 12, 1981, D6.

28. Margaret Marsh, "Men and Masculine Domesticity, 1870–1915," *American Quarterly* 40, no. 2 (June 1988): 166.

29. Bridget Kies, "Television's 'Mr. Moms': Idealizing the New Man in 1980s Domestic Sitcoms," *Feminist Media Histories* 4, no. 1 (Winter 2018): 142–70; Alice Leppert, *TV Family Values: Gender, Domestic Labor, and 1980s Sitcoms* (New Brunswick, N.J.: Rutgers University Press, 2019).

30. Kies, "Television's 'Mr. Moms.'"

31. Leppert, *TV Family Values.*

32. Peggy Watt, "Robots Roll onto the Scene," *InfoWorld,* June 11, 1984.

33. Mark Sherman, "The Home Robot Is Still in Toddler Stage," *New York Times,* March 24, 1985, sec. High Technology.

34. Michael Waldholz, "Zenith Is Promoting Its Hero One Robot for Use in Training: Heath Unit's $2,500 Product Aimed at Schools, Industry Seen Attracting Tinkerers," *Wall Street Journal,* December 2, 1982.

35. Heath, "Introducing the World's First Sophisticated Teaching Robot . . .—Advertisement," http://www.theoldrobots.com/images89/Hero-1_Ad.JPG; Heath, "Move into the World of Robotics with the First Sophisticated Teaching Robot—Hero 1—Advertisement," http://www.theoldrobots.com/images63/Hero_1.pdf.

36. David Needle, "HERO Walks, Talks, Educates, and Protects," *InfoWorld,* December 27, 1982, 4.

37. *The Impact of Robotics on Employment: Hearing before the Subcommittee on Economic Goals and Intergovernmental Policy of the Joint Economic Committee,* 98th Cong., 1st sess. (1983); *The Role of Automation and Robotics in Advancing United States Competitiveness: Hearing before the Subcommittee on Science, Research, and Technology of the Committee on Science and Technology,* 99th Cong., 1st sess. (1985).

38. Neda Atanasoski and Kalindi Vora, *Surrogate Humanity: Race, Robots, and the Politics of Technological Futures* (Durham, N.C.: Duke University Press, 2019), 2.

39. Elizabeth M. Fowler, "Careers: Learning to Handle the Robots," *New York Times,* September 17, 1980, D19.

40. William J. Broad, "Building a Robot: The Crash Course," *New York Times,* May 3, 1983, C1, C8.

41. See, for example, Mark J. Robillard, "Hero 1: Advanced Programming and Interfacing," cover image, http://www.theoldrobots.com/images98/Hero_I-k9.JPG; or "Move into the World of Robotics with Hero 1—Advertisement," http://www.theoldrobots.com/images63/Hero_1.pdf.

42. Steven M. Gelber, "Do-It-Yourself: Constructing, Repairing, and Maintaining Domestic Masculinity," *American Quarterly* 49, no. 1 (March 1997): 69.

43. Neil Shapiro, "30 Days with Hero-1," *Popular Mechanics,* September 1983.

44. Peter Owens, "At Home with . . . the Hero 1," *Popular Computing,* December 1983.

45. Heath, *Hero Jr. Owner's Guide* (Benton Harbor, MI: Heath Company, 1984), 2–3, https://archive.org/details/HEROJrOwnersGuide.

46. Heath, "Hero Jr. Entertains You with a Dynamic Personality—Advertisement," http://www.theoldrobots.com/book74/Hero_Jr_Page_4.JPG.

47. This image is archived at http://www.theoldrobots.com/book74/Hero_Jr_Page_1.JPG.

48. Heath, *Hero Jr. Owner's Guide*, 2.

49. See, for example, Heath, "Take Your Family beyond Computers—Advertisement," *Popular Science*, December 1984, 16.

50. Heath, *Hero Jr. Owner's Guide*, 3.

51. Heath, "Hero Jr. Entertains You with a Dynamic Personality."

52. *Androbot Newsletter*, no. 1 (1983), http://www.theoldrobots.com/images7/Issue_One.pdf.

53. A copy of this image can be found at http://www.theoldrobots.com/images32/Androbot15.JPG.

54. Vincent J. Schodolski, "They Walk! They Talk! Home Robots Even Fetch a Beer," *Chicago Tribune*, June 6, 1983; *Androbot Newsletter*, no. 2 (1983), http://www.theoldrobots.com/images7/Issue_Two.pdf.

55. This image is archived at https://www.robotworkshop.com/robotweb/?page_id=106.

56. *Androbot Newsletter*, no. 2.

57. Sherman, "Home Robot Is Still in Toddler Stage."

58. *Hubot Owner's Manual*, n.d., 2, http://www.theoldrobots.com/book56/HubotManual.pdf.

59. Amy Saltzman, "Robot Marketing Enters Advertising Space Age," *Adweek*, September 17, 1984.

60. This photo is archived at http://www.theoldrobots.com/images17/hu4.JPG.

61. Laura Scott Holliday, "Kitchen Technologies: Promises and Alibis, 1944–1966," *Camera Obscura* 42, no. 2 (2001): 78–131; Lynn Spigel, "Yesterday's Future, Tomorrow's Home," in *Welcome to the Dreamhouse: Popular Media and Postwar Suburbs* (Durham, N.C.: Duke University Press, 2001), 381–408.

62. Lynn Spigel, "Designing the Smart House: Posthuman Domesticity and Conspicuous Production," *European Journal of Cultural Studies* 8, no. 4 (2005): 406.

63. This image is reprinted in Jan Zimmerman, *Once upon the Future: A Woman's Guide to Tomorrow's Technology* (New York: Pandora Press, 1986), 85.

64. Fred D'Ignazio, "Is There a Robot in the House? One Family's Life with Topo," *Enter*, October 1984.

65. Stuart Diamond, "Family Life with Robot Can Be Entertaining," *New York Times*, October 18, 1984.

66. Betsy Byrne, "Rendezvous with a Robot: Your Dream 'Droid May Be Just around the Corner," *Family Computing*, March 1984.

67. Spigel, "Yesterday's Future, Tomorrow's Home"; Holliday, "Kitchen Technologies."

68. Atanasoski and Vora, *Surrogate Humanity*, 88–89.

69. Eric Zorn, "Tempo: Fathers & Sons Caring and Sharing Mark the Manly Art of Fatherhood," *Chicago Tribune,* June 20, 1982.

70. An example of the discussion of cuteness as a design tradeoff can be found in a 1983 demonstration of the Androbot, titled "Androbot Presentation" and uploaded to the Computer History Museum YouTube account on May 3, 2016, https://youtu.be/Mw4IyoKwlAc?.

71. Daniel J. Ruby, "Computerized Personal Robots: They Move, Talk, Think, and Teach You Robotics," *Popular Science,* May 1983, 100.

72. Fred D'Ignazio, "On the Road with Fred D'Ignazio: The Robot Teddy Bear," *Compute!,* January 1984, 100.

73. Androbot press release quoted in Gordon McComb, "Personal Robots," *Creative Computing,* November 1983, 198.

74. The Hero Jr. programming manual provided instructions so that users could program the robot to address the user by name as well as to customize the name the robot used to refer to itself. It also offered a list of male and female names.

75. For further discussion of masculine fantasies in sexualized representations of artificial and robot women, see Andreas Huyssen, "The Vamp and the Machine: Technology and Sexuality in Fritz Lang's *Metropolis,*" *New German Critique,* nos. 24–25 (Autumn 1981 / Winter 1982): 221–37; and Julie Wosk, *My Fair Ladies: Female Robots, Androids, and Other Artificial Eves* (New Brunswick, N.J.: Rutgers University Press, 2015). A robot representation more proximate to home robots of the 1980s can be found in the television program *Small Wonder* (1985–89), in which a father invents a robot and the family then adopts her as a daughter.

76. Anthony P. McIntyre, "Robots in Popular Culture: Labor Precarity and Machine Cute," *Flow* (blog), February 27, 2017, https://www.flowjournal.org/2017/02/robots-popular-culture/.

77. McIntyre.

78. Colin Covert, "Home Robots' Day Is Coming—but Not Yet," *Hartford Courant,* August 10, 1983.

79. Jennifer Rhee, *The Robotic Imaginary: The Human and the Price of Dehumanized Labor* (Minneapolis: University of Minnesota Press, 2018), 103–4.

80. Catherine Caudwell, Cherie Lacey, and Eduardo B. Sandoval, "The (Ir)relevance of Robot Cuteness: An Exploratory Study of Emotionally Durable Robot Design," *OzCHI '19: Proceedings of the 31st Australian Conference on Human–Computer-Interaction,* December 2, 2019, https://doi.org/10.1145/3369457.3369463.

81. "Starship Technologies: Autonomous Robot Delivery," accessed September 1, 2023, https://www.starship.xyz/; "Kiwibot Autonomous Delivery Robots, Revolutionizing the Future of Robotic Delivery," accessed September 1, 2023, https://www.kiwibot.com/; Ali Francis, "Why You Want to Pet the Food Delivery

Robot," *Bon Appétit*, October 14, 2022, https://www.bonappetit.com/story/why-food-delivery-robots-so-cute.

82. "Androbot Presentation."

83. Sianne Ngai, *Our Aesthetic Categories: Zany, Cute, Interesting* (Cambridge, Mass.: Harvard University Press, 2012), 87.

84. Lori Merish, "Cuteness and Commodity Aesthetics: Tom Thumb and Shirley Temple," in *Freakery: Cultural Spectacles of the Extraordinary Body,* ed. Rosemarie Garland Thomson (New York: NYU Press, 1996), 185–86, emphasis in original.

85. "Introducing BoB," *Boston Globe,* January 21, 1983, sec. Business.

86. Jennifer S. Light, "When Computers Were Women," *Technology and Culture* 40, no. 3 (July 1999): 462.

87. Nathan Ensmenger, "'Beards, Sandals, and Other Signs of Rugged Individualism': Masculine Culture within the Computing Professions," *Osiris* 30, no. 1 (January 2015): 38–65.

88. See, for example, Carl Quick, "Animate vs. Inanimate," *Robotics Age,* August 1984.

89. The first version of Topo was controlled by radio signals while later versions used infrared.

90. Androbot Inc., *Topo Owner's Manual, Apple II, Apple II+, and Apple IIe Computer Version,* 1983, 36, http://www.theoldrobots.com/images7/TopoManual 2.pdf.

91. Shapiro, "30 Days with Hero 1," 74, 141.

92. D'Ignazio, "On the Road."

93. Quick, "Animate vs. Inanimate," 15.

94. Ruby, "Computerized Personal Robots"; *Computer Chronicles: Robotics,* 1984, http://archive.org/details/Robotics1984.

95. Owens, "At Home with . . . the Hero 1," 166.

96. Maureen Ryan, "Why the Kitchen Computing Dream of the 80s Never Caught On," *Motherboard,* September 19, 2014, https://motherboard.vice.com/en_us/article/why-the-kitchen-computing-dream-of-the-80s-never-caught-on.

97. Hochschild, *Second Shift.*

3. "A Doll That Understands You"

1. Terms like voice recognition and speech recognition are used to describe somewhat different technologies—such as those that distinguish between different words uttered by a nonspecific speaker and those that identify the voice of a specific individual. For my purposes, I will be using the term speech recognition to describe Julie's capabilities because this is how Texas Instruments characterized this technology. As the chapter will explain, Julie was designed to recognize a limited number of specific words as spoken by a single user, who is asked to register her voice each time the doll is reset. See Gene Frantz, Jay Reimer, and Richard Wotiz, "Julie: The Application of DSP to

a Consumer Product," *Speech Technology*, September/October 1988, 82–86; Gene A. Frantz, "Toys That Talk: Two Case Studies," in *Applied Speech Technology*, ed. Ann K. Syrdal, Raymond W. Bennett, and Steven L. Greenspan (Boca Raton, Fla.: CRC Press, 1995), 487–500.

2. Miriam Formanek-Brunell, *Made to Play House: Dolls and the Commercialization of American Girlhood, 1830–1930* (New Haven, Conn.: Yale University Press, 1993).

3. "Q and A with Dr. Dylan Mulvin on Proxies: The Cultural Work of Standing In," *LSE Review of Books* (blog), August 24, 2021, https://blogs.lse.ac.uk/lsereview ofbooks/2021/08/24/q-and-a-with-dylan-mulvin-on-proxies-the-cultural -work-of-standing-in/.

4. Claudia Crowley, "'Compusuck,' Winner of 'What in the World' Contest," *InfoWorld*, October 12, 1981, 7–8.

5. Gregory Yob, "*InfoWorld* Announces 'What in the World' Results," *InfoWorld*, October 12, 1981.

6. Geraldine Carro, "What's New in Education: A Computer Boom in the Classroom," *Ladies' Home Journal*, September 1982; Dan Kaercher, "How Computers Are Changing the Classroom," *Better Homes and Gardens*, April 1983; Robert B. Kenney, "Computers Come to the Nursery," *Boston Globe*, February 19, 1984; Carol Lawson, "Computers, for Youngsters Who Can Barely Say the Word," *New York Times*, December 21, 1989.

7. Arlie Hochschild, *The Second Shift: Working Parents and the Revolution at Home* (New York: Viking, 1989); Kay Longcope, "New Notions about Parenting," *Boston Globe*, September 28, 1982.

8. Steven Mintz and Susan Kellogg, *Domestic Revolutions: A Social History of American Family Life* (New York: Free Press, 1988), 203–4.

9. Didi Moore, "For Today's 'Only Child,' 1 Isn't a Lonely Number," *Chicago Tribune*, January 18, 1981; "Size of U.S. Family Continues to Drop, Census Bureau Says," *New York Times*, June 2, 1988.

10. "The 'Latchkey Kids': They're at Home Alone while Parents Work," *Chicago Tribune*, March 6, 1983; Judy Mann, "Latchkey Kids," *Washington Post*, February 1, 1984; *Latchkey Children: Hearing before the Subcommittee on Education and Health of the Joint Economic Committee, Congress of the United States*, 100th Cong. (March 1988).

11. Alice Leppert, *TV Family Values: Gender, Domestic Labor, and 1980s Sitcoms* (New Brunswick, N.J.: Rutgers University Press, 2019), 59; Mintz and Kellogg, *Domestic Revolutions*, 223.

12. Mintz and Kellogg, *Domestic Revolutions*, 223.

13. Jeanette M. Gallagher and Judith Coche, "Hothousing: The Clinical and Educational Concerns over Pressuring Young Children," *Early Childhood Research Quarterly* 2, no. 3 (1987): 203–10; Leah Rosch, "Tempo: The Hurried Child: Pushed Too Fast into Adulthood," *Chicago Tribune*, December 11, 1984; Kathy

Seal, "Superchildren—Pushed Too Far for Their Own Good?," *Los Angeles Times*, November 20, 1987.

14. For discussion of backlash, see Susan Faludi, *Backlash: The Undeclared War against American Women* (New York: Crown, 1991). For examples of supermom discourse, see Jean Marzollo, "Don't Call Me Supermom," *Parents*, April 1984; Neala S. Schwartzberg, "'Call Me Supermom': The News Is Out—Having Lots to Do Is Good for You!," *Parents*, March 1986; Jane Marks, "The Supermom Myth, or Why Nobody's Perfect," *Parents*, February 1980; Meg Cox, "My Best Lifesavers: Supermom Tips from Working Women," *Cosmopolitan*, November 1986.

15. Gaylen Moore, "The Superbaby Myth," *Psychology Today*, June 1984.

16. Rosch, "Tempo: The Hurried Child"; Seal, "Superchildren"; Janet Elder, "The Super Baby Burnout Syndrome," *New York Times*, January 8, 1989. Reports in the 1980s switched between labeling these achieving children as "super babies" or "superbabies." In this book, I will refer to this phenomenon using a single word unless using a direct quote.

17. Carro, "What's New in Education"; Kenney, "Computers Come to the Nursery."

18. Sally Reed, "On South Shore Drive Two Professors and Their Two Preschoolers Share a Computer—in the Family Room," *Family Computing*, November 1983, 59.

19. Gene Frantz and Richard Wiggins, "The Development of 'Solid State Speech' Technology at Texas Instruments," *IEEE Acoustics, Speech, and Signal Processing Newsletter* 53, no. 1 (March 1981): 34–38.

20. Frantz, "Toys That Talk," 487.

21. Mattel, "Advertisement," *Essence*, December 1982, 97.

22. Lawson, "Computers, for Youngsters."

23. Moore, "Superbaby Myth."

24. Pamela Moreland, "Educators Divided: Super Baby: Early Start in Schooling Preschool," *Los Angeles Times*, August 31, 1985, sec. Part I; Gallagher and Coche, "Hothousing"; Anne Roark, "Parenting: Mother of Invention," *Los Angeles Times*, July 24, 1988.

25. Victoria Cain, *Schools and Screens: A Watchful History* (Cambridge, Mass.: MIT Press, 2021), 142.

26. Michael Winerip, "Rich Schools Getting Richer in Computers," *New York Times*, June 24, 1983, B1.

27. Seal, "Superchildren."

28. David Elkind, *The Hurried Child: Growing Up Too Fast Too Soon* (Reading, Mass.: Addison-Wesley, 1981); Neil Postman, *The Disappearance of Childhood* (New York: Vintage, 1994).

29. Elder, "Super Baby Burnout Syndrome"; Stephen Chapman, "Hazards of a Super Baby," *Chicago Tribune*, November 10, 1985, sec. 2; Glenn Collins, "Children: Teaching Too Much, Too Soon?," *New York Times*, November 4, 1985.

30. Seal, "Superchildren"; Elder, "Super Baby Burnout Syndrome."

31. Lawson, "Computers, for Youngsters."

32. Sally Reed, "The Concerns: What Experts Say," *Family Computing*, November 1983, 63; Lawson, "Computers, for Youngsters."

33. Fred D'Ignazio, "The World Inside the Computer: Superbaby Meets the Computer," *Compute!*, July 1983, 150.

34. Moore, "Superbaby Myth," 6.

35. Collins, "Children."

36. Lynn Langway, "Bringing Up Superbaby," *Newsweek*, March 28, 1983.

37. For example, one of Sherry Turkle's chapters on children and computing opens with an anecdote describing a young blonde girl in a pinafore, perhaps to emphasize the dissonance between this image of white girlhood innocence and the high-tech computer. See *The Second Self: Computers and the Human Spirit* (New York: Simon and Schuster, 1984), chapter 3.

38. Morgan G. Ames, *The Charisma Machine: The Life, Death, and Legacy of One Laptop per Child* (Cambridge, Mass.: MIT Press, 2019), 31.

39. "Girls Get Advice on Dealing with Technology," *New York Times*, May 15, 1983.

40. Juliet F. Brudney, "Girls: Think Computers," *Boston Globe*, April 19, 1984; Joseph F. Sullivan, "Schools Urge Girls to Use Computers," *New York Times*, May 5, 1986; Carol Kleiman, "The Work Place: Girls Clubs Pushes Leap into Technological Society," *Hartford Courant*, March 24, 1986.

41. An example of this can be found in "Rhiannon Software Advertisement," *Family Computing*, December 1984, 91.

42. Michael Z. Newman, *Atari Age: The Emergence of Video Games in America* (Cambridge, Mass.: MIT Press, 2017), 136–37.

43. Teddy Ruxpin from Worlds of Wonder was a standout toy in 1985. See James Bates, "Teddy in a Tumult: Problems of Toy's Producer Leave Its Creator in a Bind," *Los Angeles Times*, January 19, 1988. Teddy Ruxpin did not feature digitally synthesized speech. This bear was based on a patent from Ken Forsse, who began his career working on animatronics at Disney. Teddy Ruxpin's movement synced with speech sounds from specially encoded cassette tapes.

44. Formanek-Brunell, *Made to Play House*.

45. Formanek-Brunell; Ann DuCille, "Toy Theory: Black Barbie and the Deep Play of Difference," in *The Consumer Society Reader*, ed. Juliet B. Schor and Douglas B. Holt (New York: New Press, 2000), 259–78; Erica Rand, *Barbie's Queer Accessories* (Durham, N.C.: Duke University Press, 1995); Robin Bernstein, *Racial Innocence: Performing American Childhood from Slavery to Civil Rights* (New York: NYU Press, 2011).

46. For technical details about Baby Talk's speech technology, see Jay Smith III, "Speech Products and Toys: Christmas Past, Christmas Present, and Christmas Future," *Official Proceedings of Speech Tech*, 1988, 16.

47. Denise Gellene, "$100 Dolls Are the Talk of Toy Makers," *Los Angeles Times*, November 22, 1987; Associated Press, "Galoob Toys Recognized as an Industry Leader," *Hartford Courant*, March 28, 1987.

48. Lewis Galoob Toys, *Helping Your Child Learn with Baby Talk,* author's collection.
49. Lewis Galoob Toys.
50. For a discussion of talking doll mechanisms over the twentieth century, see Meredith Bak, "Between Technology and Toy: The Talking Doll as Abject Artifact," in *Abjection Incorporated: Mediating the Politics of Pleasure and Violence,* ed. Maggie Hennefeld and Nicholas Sammond (Durham, N.C.: Duke University Press, 2020), 164–84.
51. Lewis Galoob Toys, *Helping Your Child Learn with Baby Talk.*
52. Lewis Galoob Toys.
53. Lewis Galoob Toys.
54. Lewis Galoob Toys trade catalog, 1987, from the Stephen and Diane Olin Toy Catalog Collection, the Strong, Rochester, N.Y., 32.
55. Ralph H. Baer, "Ground Rules," January 17, 1986, folder 8, box 5, Ralph H. Baer Papers, Brian Sutton-Smith Library and Archives of Play, the Strong, Rochester, N.Y.
56. Jim Fuller, "Twin Cities Kids Helping Baby-Sit 100 New Arrivals," *Minneapolis Star and Tribune,* July 23, 1986.
57. Lewis Galoob Toys, *Helping Your Child Learn with Baby Talk.*
58. Frantz, "Toys That Talk," 496.
59. Frantz, 498.
60. Gellene, "$100 Dolls Are the Talk of Toy Makers."
61. Eugenia Gonzalez, "'I Sometimes Think She Is a Spy on All My Actions': Dolls, Girls, and Disciplinary Surveillance in the Nineteenth-Century Doll Tale," *Children's Literature* 39 (2011): 33–57.
62. Angela McRobbie, "Top Girls? Young Women and the Post-feminist Sexual Contract," *Cultural Studies* 21, nos. 4–5 (July/September 2007): 722.
63. McRobbie, 721.
64. Anita Harris, *Future Girl: Young Women in the Twenty-First Century* (New York: Routledge, 2004).
65. Ann duCille, "Dyes and Dolls: Multicultural Barbie and the Merchandising of Difference," in *The Black Studies Reader,* ed. Jacqueline Bobo, Cynthia Hudley, and Claudine Michel (New York: Routledge, 2004), 267.
66. One effort some scholars have deemed more successful was produced by Shindana Toys in 1968. The company's Baby Nancy doll was said to be modeled on the faces of specific local girls from the Watts neighborhood in Los Angeles, where the doll was designed and manufactured and where its target audience resided. For research on Shindana, see David Crittendon, Yolanda Hester, and Rob Goldberg, "The Legacy of Shindana Toys: Black Play and Black Power, an Interview with David Crittendon, Yolanda Hester, and Rob Goldberg," *American Journal of Play* 13, nos. 2–3 (Winter/Spring 2021): 135–46.
67. Ellen Goodman, "Toys Come Alive: They Talk—but They Still Don't Listen," *Los Angeles Times,* December 16, 1986, B5.

68. Linda Wells, "The Toys at the Top of Children's Gift Lists," *New York Times,* November 29, 1986; Deborah Light, "Animatronics, the New Wave Toys for Christmas," *Sydney Morning Herald,* November 5, 1986.

69. Smith, "Speech Products and Toys," 16.

70. For a discussion of the relationship between girls and dolls like Baby Talk, Julie, and Jill as a form of cybernetic interaction, see Reem Hilu, "Girl Talk and Girl Tech: Computer Talking Dolls and the Sounds of Girls' Play," *Velvet Light Trap* 78 (2016): 4–21.

71. Lewis Galoob Toys, *Helping Your Child Learn with Baby Talk.*

72. Kenneth J. Curran, Interactive Talking Toy, U.S. Patent 4,923,428, filed May 5, 1988, and issued May 8, 1990.

73. Frieda Rebelski quoted in Ronald Rosenberg, "High-Tech Toys Finding More and More to Say; Stuffed with Microelectronics, Dolls Can Talk, Act and React," *Boston Globe,* December 22, 1986, sec. Science and Technology. For other examples of this discourse see Goodman, "Toys Come Alive"; Sylvia L. Wilson, "'TV Toys': Debut and Debate," *New York Times,* February 9, 1987, sec. Business Day, D3; Kenneth R. Clark, "Bellicose Toys Don't Look Like Child's Play," *Chicago Tribune,* February 15, 1987, D3; and David Streitfeld, "Dolls That Yakety-Yak: Trends," *Washington Post,* September 18, 1987, sec. Style, D5. This concern was not new to computer dolls. For similar anxieties regarding earlier talking dolls such as Chatty Cathy, see Robert Wallace, "Look! Its Nose Runs!," *Life,* December 18, 1950; and David Moller, "Christmas and Freud," *Saturday Evening Post,* December 7, 1963, 38.

74. "Spectator Toys Called Some of the Worst for Kids," *Star Tribune,* November 7, 1988.

75. Betty Holcomb, "The Brave New World of Interactive Toys: They Walk, They Wisecrack, but Do They Enrich a Child's Play?," *Working Mother,* October 1987.

76. Clark, "Bellicose Toys Don't Look Like Child's Play."

77. For example, see Richard deCordova, "The Mickey in Macy's Window: Childhood, Consumerism, and Disney Animation," in *Disney Discourse: Producing the Magic Kingdom,* ed. Eric Smoodin (New York: Routledge, 1994), 203–13; Richard deCordova, "Child Rearing Advice and the Moral Regulation of Children's Movie-Going," *Quarterly Review of Film and Video* 15, no. 4 (1995): 99–109; Nicholas Sammond, *Babes in Tomorrowland: Walt Disney and the Making of the American Child, 1930–1960* (Durham, N.C.: Duke University Press, 2005); Lisa Jacobson, *Raising Consumers: Children and the American Mass Market in the Early Twentieth Century* (New York: Columbia University Press, 2004); Lynn Spigel, "Seducing the Innocent: Childhood and Television in Postwar America," in *Welcome to the Dreamhouse: Popular Media and Postwar Suburbs* (Durham, N.C.: Duke University Press, 2001), 185–218; and Jacob Smith, "Turntable Jr.," in *Spoken Word: Postwar American Phonograph Culture* (Berkeley: University of California Press, 2011), 13–48.

78. Jacqueline Rose, *The Case of Peter Pan, Or the Impossibility of Children's Fiction* (Philadelphia: University of Pennsylvania Press, 1984).

79. Heather Hendershot, *Saturday Morning Censors: Television Regulation before the V-Chip* (Durham, N.C.: Duke University Press, 1998), 95–135; Ellen Seiter, *Sold Separately: Children and Parents in Consumer Culture* (New Brunswick, N.J.: Rutgers University Press, 1995).

80. Justine Cassell and Henry Jenkins, eds., *From Barbie to Mortal Kombat: Gender and Computer Games* (Cambridge, Mass.: MIT Press, 1998); Marsha Kinder, *Playing with Power in Movies, Television, and Video Games: From Muppet Babies to Teenage Mutant Ninja Turtles* (Berkeley: University of California Press, 1991), 87–120.

81. Seiter, *Sold Separately*, 145–71.

82. Ralph H. Baer, "Interactive Video-Nested Data Control System," 1986, folder 8, box 5, Ralph H. Baer Papers, Brian Sutton-Smith Library and Archives of Play, the Strong, Rochester, New York. The first category includes the types of games played on the Magnavox Odyssey home video game console Baer was credited with inventing.

83. Baer.

84. Ralph H. Baer, Video-Based Instructional and Entertainment System Using Animated Figure, U.S. Patent 4,846,693, filed December 1, 1988, and issued July 11, 1989.

85. Bernadette Flynn, "Geography of the Digital Hearth," *Information, Communication and Society* 6, no. 4 (2003): 551–76.

86. Newman, *Atari Age*, 76–77.

87. Henry Jenkins, "'Complete Freedom of Movement': Video Games as Gendered Play Spaces," in *From Barbie to Mortal Kombat: Gender and Computer Games*, ed. Justine Cassell and Henry Jenkins (Cambridge, Mass.: MIT Press, 1998), 262–97.

88. Cassell and Jenkins, *From Barbie to Mortal Kombat*. The focus of the girls' game movement on screen-based computer games like *Barbie Fashion Designer* and *The Sims* franchise has obscured the importance of other types of girls' computer use. There has been some scholarship on girls' use of technologies like Tamagotchi, virtual pets, and cell phones, but these are rarely considered as significant examples of girls' computer use despite the fact that these are small, portable computers. See Anne Allison, "Tamagotchi: The Prosthetics of Presence," in *Millennial Monsters: Japanese Toys and the Global Imagination* (Berkeley: University of California Press, 2006), 163–91; Ellen Seiter, "Virtual Pets: Devouring the Children's Market," in *The Internet Playground: Children's Access, Entertainment and Mis-education* (New York: Peter Lang, 2005), 83–100; Leslie Regan Shade, "Feminizing the Mobile: Gender Scripting of Mobiles in North America," *Continuum: Journal of Media & Cultural Studies* 21, no. 2 (June 2007): 179–89.

4. Sex and the Singles Game

1. Microillusions, "Microillusions Catalog," n.d., https://archive.org/details/vg museum_miscgame_microillusions-catalog.

2. Microillusions, "*Romantic Encounters at the Dome* Manual," n.d., author's collection.

3. Jean Dietz, "Logged On, Turned Off: Some Are Afraid of Computers, Some Are Obsessed with Them," *Boston Globe*, February 6, 1984; Peter H. Lewis, "When Machines Spawn Obsession," *New York Times*, November 13, 1988.

4. Carolyn Bronstein, "Clashing at Barnard's Gates: Understanding the Origins of the Pornography Problem in the Modern American Women's Movement," in *New Views on Pornography: Sexuality, Politics, and the Law*, ed. Lynn Comella and Shira Tarrant (Santa Barbara, Calif.: Praeger, 2015), 57–76; Barbara Ehrenreich, "A Feminist's View of the New Man," *New York Times*, May 20, 1984, sec. Magazine; Deborah Laake, "A World Full of Wormboys," *San Diego Reader*, December 8, 1983; Curt Suplee, "Dawn of the Wimp: You Asked for It: The Creature from Beyond Sincere," *Washington Post*, September 10, 1984.

5. Julia Havas and Maria Sulimma, "Through the Gaps of My Fingers: Genre, Femininity, and Cringe Aesthetics in Dramedy Television," *Television and New Media* 21, no. 1 (2020): 75–94.

6. For examples of the discussion of masculine ideals, see Anastasia Salter and Bridget Blodgett, *Toxic Geek Masculinity in Media: Sexism, Trolling, and Identity Policing* (New York: Palgrave MacMillan, 2017), chapter 4; and Derek A. Burrill, *Die Tryin': Videogames, Masculinity, Culture* (New York: Peter Lang, 2008). For examples of scholarship that discusses critical or denaturalized masculine representation, see Ewan Kirkland, "Masculinity in Video Games: The Gendered Gameplay of *Silent Hill*," *Camera Obscura* 24, no. 2 (2009): 160–83; Ian Bryce Jones, "Do the Locomotion: Obstinate Avatars, Dehiscent Performances, and the Rise of the Comedic Video Game," *Velvet Light Trap* 77 (Spring 2016): 86–99; and Bo Ruberg, *Video Games Have Always Been Queer* (New York: NYU Press, 2019), chapter 3.

7. One exception that will be discussed later in the chapter is Mia Consalvo, "Hot Dates and Fairy-Tale Romances: Studying Sexuality in Video Games," in *The Video Game Theory Reader*, ed. Mark J. P. Wolf and Bernard Perron (New York: Routledge, 2003), 171–94.

8. Duncan Fyfe, "Was *Leisure Suit Larry* Really an Accomplice in Early Banking Cyberattacks?," *Vice* (blog), April 8, 2020, https://www.vice.com/en/article/epg9be/did-sierra-create-leisure-suit-larry-virus-stop-piracy.

9. Maurice K. Byte, Steve Carter, and Josh Levine, *How to Make Love to a Computer* (New York: Pocket Books, 1984), 13.

10. Sherry Turkle, *The Second Self: Computers and the Human Spirit* (New York: Simon and Schuster, 1984).

11. Georgia Dullea, "New Marital Stress: The Computer Complex," *New York Times,* January 10, 1983; Dietz, "Logged On, Turned Off"; Lewis, "When Machines Spawn Obsession."

12. Ehrenreich, "Feminist's View of the New Man."

13. Barbara Ehrenreich, *The Hearts of Men: American Dreams and the Flight from Commitment* (Garden City, N.Y.: Doubleday, 1983), 12.

14. Ehrenreich, "Feminist's View of the New Man."

15. Laake, "World Full of Wormboys."

16. Michael Kimmel, *Manhood in America: A Cultural History* (New York: Oxford University Press, 1996), 292.

17. Notably, Alan Alda served as a spokesperson for Atari Computers, another instance in which the New Man functioned to help situate computers in the home. Bruce Feirstein, *Real Men Don't Eat Quiche* (New York: Pocket Books, 1982), 9, emphasis in original.

18. Bronstein, "Clashing at Barnard's Gates," 60–62.

19. Bronstein.

20. R. W. Connell, *Masculinities,* 2nd ed. (Berkeley: University of California Press, 2005), 55–56.

21. Lori Kendall, "Nerd Nation: Images of Nerds in U.S. Popular Culture," *International Journal of Cultural Studies* 2, no. 2 (1999): 260–83.

22. Kendall; Ron Eglash, "Race, Sex, and Nerds: From Black Geeks to Asian American Hipsters," *Social Text* 20, no. 2 (Summer 2002): 49–64.

23. Dan Sheridan, "Computer Games May Help Business," *Crain's Chicago Business,* November 28, 1988; Barbara T. Roessner, "Software Porn Brings Rape to the Office," *Hartford Courant,* December 3, 1988; Patt Morrison, "Fun and Games or Compu-Porn?," *Los Angeles Times,* November 14, 1988, sec. V.

24. Elana Levine, *Wallowing in Sex: The New Sexual Culture of 1970s American Television* (Durham, N.C.: Duke University Press, 2007).

25. Peter Alilunas, *Smutty Little Movies: The Creation and Regulation of Adult Video* (Oakland: University of California Press, 2016), 6.

26. Luke Stadel, "Cable, Pornography, and the Reinvention of Television, 1982–1989," *Cinema Journal* 53, no. 3 (Spring 2014): 61, 74.

27. Alilunas, *Smutty Little Movies,* 164.

28. Alilunas, 158–200.

29. Jimmy Maher, "Sex Comes to the Micros," *Digital Antiquarian* (blog), February 27, 2012, https://www.filfre.net/2012/02/sex-comes-to-the-micros/.

30. Jimmy Maher reports that fifty thousand copies of *Softporn* were sold, a large number at the time. As many have pointed out, this would have made up a sizable portion of the computer owning public at this early date. "Softporn" *Digital Antiquarian* (blog), February 29, 2012, https://www.filfre.net/2012/02/softporn/.

31. Laine Nooney, "The Odd History of the First Erotic Computer Game," *The Atlantic*, December 2, 2014, https://www.theatlantic.com/technology/archive/2014/12/the-odd-history-of-the-first-erotic-computer-game/383114/.

32. Matthew Thomas Payne and Peter Alilunas, "Regulating the Desire Machine: *Custer's Revenge* and 8-Bit Atari Porn Video Games," *Television and New Media* 17, no. 1 (2016): 83.

33. Payne and Alilunas.

34. Bronstein, "Clashing at Barnard's Gates."

35. Payne and Alilunas, "Regulating the Desire Machine," 84.

36. Morrison, "Fun and Games or Compu-Porn?," F1.

37. Roessner, "Software Porn Brings Rape to the Office," C1.

38. Morrison, "Fun and Games or Compu-Porn?"

39. John Williams, "Goodbye 'G' Ratings: The New Wave of Adult Entertainment Software [Part 1]," *Computer Gaming World,* August/September 1987, 30.

40. Adrienne Shaw, "*Leisure Suit Larry:* LGBTQ Representation," in *How to Play Video Games,* ed. Matthew Thomas Payne and Nina B. Huntemann (New York: NYU Press, 2019), 110–17; Jimmy Maher, "*Leisure Suit Larry in the Land of the Lounge Lizards,*" *Digital Antiquarian* (blog), August 15, 2015, https://www.filfre.net/2015/08/leisure-suit-larry-in-the-land-of-the-lounge-lizards/.

41. Williams, "Goodbye 'G' Ratings [Part 1]," 30–31.

42. Williams, 31.

43. Williams, 31.

44. John Williams, "Goodbye 'G' Ratings: The New Wave of Adult Entertainment Software [Part 3]," *Computer Gaming World,* January 1988, 49.

45. Williams, "Goodbye 'G' Ratings [Part 1]," 31.

46. John Williams, "Goodbye 'G' Ratings: The New Wave of Adult Entertainment Software [Part 2]," *Computer Gaming World,* October 1987, 52–53.

47. Philip Jong, "Al Lowe: iBase Entertainment," *Adventure Classic Gaming,* May 17, 2006, http://www.adventureclassicgaming.com/index.php/site/interviews/199/.

48. Havas and Sulimma, "Through the Gaps of My Fingers," 82–83.

49. Josef Nguyen and Bo Ruberg, "Challenges of Designing Consent: Consent Mechanics in Video Games as Models for Interactive User Agency," in *CHI '20: Proceedings of the 2020 CHI Conference on Human Factors in Computing Systems,* April 2020, 4.

50. Sierra, "*Leisure Suit Larry in the Land of the Lounge Lizards*—Instruction Manual," https://archive.org/details/Leisure_Suit_Larry_-_Manual.

51. Maher, "*Leisure Suit Larry in the Land of the Lounge Lizards.*"

52. Constance Penley, "Crackers and Whackers: The White Trashing of Porn," in *Pornography: Film and Culture,* ed. Peter Lehman (New Brunswick, N.J.: Rutgers University Press, 2006), 108, 105–6.

53. Shaw, "*Leisure Suit Larry*," 116.

54. Sierra, "Sierra Catalog—Fall 1988," n.d., https://archive.org/details/Sierra_Catalog_Fall_1988_1988_Sierra-On_Line.

55. Keith McCandless, "How Not to Meet Women," *Macworld,* December 1987, 142.

56. Sierra, "*Leisure Suit Larry* Collection Series Manual," n.d., https://archive.org/details/Leisure_Suit_Larry_Collection_-_Manual.

57. Maher, "*Leisure Suit Larry*."

58. Shaw, "*Leisure Suit Larry*," 113–14.

59. Shaw, 115–16.

60. Anastasia Salter, "Plundered Hearts: Infocom, Romance, and the History of Feminist Game Design," *Feminist Media Histories* 6, no. 1 (2020): 84.

61. Salter, 84.

62. Roy Wagner, "*Leisure Suit Larry:* The Slimier Things of Life," *Computer Gaming World,* November 1987, 44.

63. Kenneth E. Schaefer, "Review: *Leisure Suit Larry in the Land of the Lounge Lizards,*" *Amazing Computing,* February 1988, 58.

64. McCandless, "How Not to Meet Women," 142.

65. Arnie Katz, "Review: *Leisure Suit Larry in the Land of the Lounge Lizards,*" *ST-Log,* June 1988, 84.

66. Jong, "Al Lowe."

67. Microillusions, "Microillusions Catalog."

68. Veronica Thomas, phone conversation with author, August 18, 2023.

69. See, for example, Salter and Blodgett, *Toxic Geek Masculinity in Media;* Carly A. Kocurek, *Coin-Operated Americans: Rebooting Boyhood at the Video Game Arcade* (Minneapolis: University of Minnesota Press, 2015); Megan Condis, *Gaming Masculinity: Trolls, Fake Geeks, and the Gendered Battle for Online Culture* (Iowa City: University of Iowa Press, 2018); Burrill, *Die Tryin'.*

70. Salter and Blodgett, *Toxic Geek Masculinity in Media,* 85.

71. Ran Almog and Danny Kaplan, "The Nerd and His Discontent: The Seduction Community and the Logic of the Game as a Geeky Solution to the Challenges of Young Masculinity," *Men and Masculinities* 20, no. 1 (2017): 28.

72. Kirkland, "Masculinity in Video Games," 166, 175.

73. Jones, "Do the Locomotion," 88.

74. Ruberg, *Video Games Have Always Been Queer,* 85.

75. Consalvo, "Hot Dates and Fairy-Tale Romances," 173–74.

76. Consalvo, 178–79.

Coda

1. A recording of the commercial, titled "Apple Personal Computers—'Average Homemaker' with Dick Cavett (Commercial, 1981)," was uploaded to the

Museum of Classic Chicago Television YouTube channel on December 28, 2022, https://www.youtube.com/watch?v=eJehPaewjqA.

2. This 1983 demonstration of the Androbot, titled "Androbot Presentation," is available from the Computer History Museum account on YouTube, uploaded May 3, 2016, https://www.youtube.com/watch?v=Mw4IyoKwlAc.

3. Lynn Spigel, *Make Room for TV: Television and the Family Ideal in Postwar America* (Chicago: University of Chicago Press, 1992); Keir Keightley, "'Turn It Down!' She Shrieked: Gender, Domestic Space, and High Fidelity, 1948–59," *Popular Music* 15, no. 2 (May 1996): 149–77; Haidee Wasson, "Electric Homes! Automatic Movies! Efficient Entertainment! 16mm and Cinema's Domestication in the 1920s," *Cinema Journal* 48, no. 4 (Summer 2009): 1–21; Jacob Smith, *Spoken Word: Postwar American Phonograph Cultures* (Berkeley: University of California Press, 2011); Carolyn Marvin, *When Old Technologies Were New: Thinking about Electric Communication in the Late Nineteenth Century* (New York: Oxford University Press, 1988); Jonathan Sterne, *The Audible Past: Cultural Origins of Sound Reproduction* (Durham, N.C.: Duke University Press, 2003).

4. Jennifer S. Light, "When Computers Were Women," *Technology and Culture* 40, no. 3 (July 1999): 455–83; Nathan Ensmenger, *The Computer Boys Take Over: Computers, Programmers, and the Politics of Technical Expertise* (Cambridge, Mass.: MIT Press, 2010); Mar Hicks, *Programmed Inequality: How Britain Discarded Women Technologists and Lost Its Edge in Computing* (Cambridge, Mass.: MIT Press, 2017).

5. Laine Nooney, "A Pedestal, a Table, a Love Letter: Archaeologies of Gender in Videogame History," *Game Studies* 13, no. 2 (December 2013), http://gamestudies.org/1302/articles/nooney.

6. "Playmates Toys," accessed September 4, 2023, https://www.encyclopedia.com/books/politics-and-business-magazines/playmates-toys.

7. Scott Mace, "Atari's Founder Tells Lessons, Plans," *PCWorld,* September 11, 2003.

8. Brenda Salinas, "In Sync: Is Sharing Your Online Calendar a Relationship Milestone?," *NPR, All Tech Considered* (blog), March 19, 2016, https://www.npr.org/sections/alltechconsidered/2016/03/19/471019632/in-sync-is-sharing-your-online-calendar-a-relationship-milestone.

9. Emily A. Vogels and Monica Anderson, "Dating and Relationships in the Digital Age," *Pew Research Center: Internet, Science, and Tech* (blog), May 8, 2020, https://www.pewresearch.org/internet/2020/05/08/dating-and-relationships-in-the-digital-age/.

10. Ja-Young Sung et al., "'My Roomba Is Rambo': Intimate Home Appliances," in *UbiComp 2007: Ubiquitous Computing,* ed. John Krumm et al. (Berlin: Springer, 2007), 151; Jodi Forlizzi and Carl DiSalvo, "Service Robots in the Domestic Environment: A Study of the Roomba Vacuum in the Home," *HRI*

'06: *Proceedings of the 1st ACM SIGCHI/SIGART Conference on Human-Robot Interaction,* March 2006, 258–65.

11. Notably, although Amazon heavily promoted the Astro robot for home use in their promotional Devices and Services event in fall 2021, it was reported that the robot was not mentioned at all during the same event the following year. See Todd Bishop, "Where's Astro? Amazon Addresses Home Robot's Absence from Annual Devices Event," *GeekWire,* September 22, 2023, https://www.geekwire.com/2023/wheres-astro-amazon-addresses-home-robots-absence-from-annual-devices-event/.

12. Amazon, "Introducing Amazon Astro-Household Robot for Home Monitoring, with Alexa," uploaded September 28, 2021, YouTube video, https://youtu.be/sj1t3msy8dc?.

13. Thao Phan, "Amazon Echo and the Aesthetics of Whiteness," *Catalyst: Feminism, Theory, Technoscience* 5, no. 1 (2019): 1–39; Neda Atanasoski and Kalindi Vora, *Surrogate Humanity: Race, Robots, and the Politics of Technological Futures* (Durham, N.C.: Duke University Press, 2019).

14. Amazon Alexa, "Echo Dot Kids Edition," uploaded Apr. 25, 2018, YouTube video, https://youtu.be/jNdZAgij-K0?.

15. Natasha Singer, "Amazon to Pay $25 Million to Settle Children's Privacy Charges," *New York Times,* May 31, 2023, https://www.nytimes.com/2023/05/31/technology/amazon-25-million-childrens-privacy.html; Leah Nylen and Matt Day, "Amazon Faces FTC Complaint Alexa Illegally Collected Kids' Data," *Bloomberg,* March 31, 2023, https://www.bloomberg.com/news/articles/2023-03-31/amazon-faces-ftc-complaint-alexa-illegally-collected-kids-data; Sam Biddle, "Experts Say Keep Amazon's Alexa Away from Your Kids," *The Intercept,* May 11, 2018, https://theintercept.com/2018/05/11/experts-say-keep-amazons-alexa-away-from-your-kids/; Amelia Hill, "Voice Assistants Could 'Hinder Children's Social and Cognitive Development,'" *The Guardian,* September 28, 2022, https://www.theguardian.com/technology/2022/sep/28/voice-assistants-could-hinder-childrens-social-and-cognitive-development.

16. Ran Almog and Danny Kaplan, "The Nerd and His Discontent: The Seduction Community and the Logic of the Game as a Geeky Solution to the Challenges of Young Masculinity," *Men and Masculinities* 20, no. 1 (2017): 27–48.

17. Steam Store, "Super Seducer: How to Talk to Girls," n.d., https://store.steampowered.com/app/695920/Super_Seducer__How_to_Talk_to_Girls/.

18. This is part of a larger project of interrogating the long history of social media. See Rebecca Wanzo and Reem Hilu, "Editors' Introduction: The Long History of Social Media," *Feminist Media Histories* 10, no. 1 (Winter 2024): 1–16.

19. Reviews of Coral on Apple's App Store suggest that these attempts at inclusivity are not especially successful.

INDEX

adult games. *See* video games: adult
adult videos, 23, 140–42. *See also*
    pornography/porn
advertising: of Apple II as personal
    appliance, 188n40; computer, 66; for
    Echo Dot Kids, 177; for Hero 1, 75,
    77; for *Interlude*, 38, 42, 47, 58; for
    IntraCourse, 59–60; for *Lovers or
    Strangers*, 29, 47, 50; personal robot,
    82–83, 85–86; for romance software,
    39, 58; for talking dolls, 111
aesthetics: of *Leisure Suit Larry in the
    Land of the Lounge Lizards*, 151; of
    machine cuteness, 88; of neoliberal
    girlhood, 122; of romance novel
    covers, 29, 30 (fig.), 47, 50, 173. *See
    also* cringe aesthetics
Alda, Alan, 17, 138, 208n17
Alexa, 1–2, 13, 175, 177
Alilunas, Peter, 141–42
Almog, Ran, 167, 177
Alpine Software, 49–50. *See also* Crane,
    Stanley; *Lovers or Strangers*
Altair, 15, 188n37
Amazon Echo, 1–2, 13, 175
Androbot, 25, 63, 81–82, 87–90, 93,
    173, 175–76; cuteness and, 88–89,
    199n70; promotion of robots, 64, 74,

84. *See also* BoB; Bushnell, Nolan;
    Frisina, Tom; Topo
antipornography feminists, 132–33,
    138, 141–43, 151
anxiety, 11, 18, 107; *Leisure Suit Larry
    in the Land of the Lounge Lizards*
    and, 152; men's sexual performance,
    135; parents and, 104, 118; sexuality
    and, 133, 148; tech, 106
Apple Computer, 4, 15, 17, 68, 171; App
    Store, 212n19; *MacPlaymate* and, 144
Apple II, 3, 15–16, 66–67, 171–72,
    188n40; children and, 105; history
    of, 8; *Interlude* and, 41; *Lovers or
    Strangers* and, 29, 47; Topo and, 81,
    93
appliances, 9, 129; Apple and, 171,
    188n40; automation of, 19; comput-
    ers as, 7–8; domestic, 7, 64–65, 84;
    household, 17; kitchen, 85; robots
    as, 81, 83, 173
Atanasoski, Neda, 76, 86, 177
Atari, 16–17, 81, 99, 175; Computers,
    208n17; 600XL, 20; VCS, 81, 141;
    video game console, 83
automation, 72, 77, 89; home, 17;
    industrial, 68–69, 75–76; of the
    workplace, 14

Baby Talk, 100, 108, 110–16, 120–29, 175, 177; Black, 122–23; cybernetic interaction and, 205n70; sound activation and, 112, 114–15, 121; speech recognition and, 112; speech technology of, 203n46

Baer, Ralph, 115, 128–29, 206n82

Black, Michael, L, 7–8

Black civil rights, 21–22, 139

Blodgett, Bridget, 167, 207n6, 210n69

BoB, 72–73, 81–82, 87–90

bondage, 40, 44

boy culture, 129–30

boys, 12, 108, 120, 127, 142; play cultures of, 102; teenage, 146

Brothers, Joyce, 54 (fig.), 55

Bushnell, Nolan, 81, 90, 175

business computers, 16, 136

Byte, 4, 5 (fig.), 15–16, 186n8, 195n8

Cavett, Dick, 17, 171

Ceruzzi, Paul, 14, 188n37

childcare, 7, 22–23, 33, 74, 104; dolls and, 100–101; fathers and, 70–71, 95; lobbyist organization, 116; masculine domesticity and, 77; robots and, 65, 81–82, 94, 96–98

child-rearing, 23, 26, 68; adult games and, 134; anxieties about, 104, 123; dolls and, 100–101, 119–20, 123–24; maternal, 24, 128; men's participation in, 25, 71, 73; play and, 126; practices, 101, 127; robots and, 73, 81, 94, 96; values, 22

children, 4, 12, 18–23, 25, 100, 102–9, 177, 203n37; adult games and, 134, 141, 146; dolls and, 101, 111–16, 118–19, 121–27; over-programmed, 104, 107; participant fatherhood and, 64–68, 70–72, 74–75, 86; robots and, 64–67, 72–75, 78, 80–82, 84–88, 96–97, 176; robots

as, 94, 96, 176; women with, 35, 69, 103. *See also* hothousing; play; superbabies

class, 12; disparity, 107; new masculinity and, 137; status, 69

Coleco, 16; Talking Cabbage Patch dolls, 110

Commodore, 15, 17; Amiga, 131, 158; PET, 3, 7, 15; 64, 189n50; TRS-80, 3, 7, 15, 41; VIC-20, 189n50

communication, 23, 31, 34, 148, 176; computer as device for, 10; computer widows and, 36; difficulties in, 33; *Interlude* and, 41–44, 46–47; *Intracourse* and, 55, 58; *Lovers or Strangers* and, 49–52, 58; between robot and computer, 93; style of computers, 19

companionate bonds, 98, 172

companionate computing, 2, 8–11, 24, 175

companionate family, 19–21, 24, 27, 65, 179; adult games and, 140; computer culture and, 172; computer talking dolls and, 102; domestic technologies and, 86; fatherhood in, 74, 82; *Leisure Suit Larry in the Land of the Lounge Lizards* and, 145–46; media technologies and, 3; men and, 68, 71; relations, 6, 14; relationships, 17; robots and, 86, 92, 97; roles in, 11; romance software and, 32, 61; women working outside the home and, 22

companionate ideals, 21, 133, 179

companionate relationships, 2, 7–8, 11, 24, 26–27, 160, 172–73; digital media and, 178; Topo and, 85

companionship, 2, 23, 84; in the home, 20; marriage and, 21; robots and, 88; social, 100, 111–12; talking dolls and, 100–101, 119, 121

competition, 17, 32, 139
computer addiction, 18, 132
computer games. *See* video games
computer history, 3, 7, 38, 174
computer industry, 13, 16–18, 20, 23;
  adult games and, 132; British, 174;
  male-dominated, 46; masculine, 39
computerization, 68–69, 72; of everyday life, 64
computer phobia, 6, 18, 37
computer programs, 6, 19, 47, 95–96,
  134, 143; children and, 105; companionate relationality and, 24–25;
  database, 7; domestic sexuality and,
  38; robots and, 78–80, 93; women
  and, 3. See also *Interlude*; *Intra-
  Course*; *Lovers or Strangers*; romance
  software; video games
computer widows, 4, 18, 31–33, 35–38,
  186n9, 191n2
computing. *See* companionate computing; home computing; personal
  computing
computing technology, 2–3, 14, 23, 99,
  109, 141; educational, 105; in the
  home, 13, 61; robots as, 26. *See also*
  hardware; software
confession, 41, 51–52
Consalvo, Mia, 168, 207n7
consumers, 2, 15–17, 50, 73, 135–36;
  Baby Talk and, 116; Black, 123;
  Heathkits and, 75; *Leisure Suit Larry
  in the Land of the Lounge Lizards*
  and, 147; male, 72; as maternal, 89;
Coral, 176, 178–79, 212n19
coupled relationships, 23, 41, 44,
  46–47, 56–57, 176
couples, 11, 23–25, 34–37, 61–62; dual-
  career, 33, 35; heterosexual, 31, 34,
  36, 39, 42, 58; *Interlude* and, 41–47,
  58, 179; *IntraCourse* and, 56, 59;
  Lovers or Strangers and, 29, 31, 51,

59–60, 179; married, 22; romance
  software and, 31–33, 37, 39, 58, 176,
  178; romantic, 39, 176
coupling, 35, 51, 58, 61–62, 173;
  heterosexual, 34, 41; imperative, 57
Crane, Stanley, 49–50, 59
Creative Computing, 29, 30 (fig.), 60
cringe, 133, 144–45, 148–52, 155,
  157–58, 161, 163–64, 168, 178. See
  also *Leisure Suit Larry in the Land
  of the Lounge Lizards*; *Romantic
  Encounters at the Dome*
cringe aesthetics: in *Leisure Suit Larry
  in the Land of the Lounge Lizards*,
  26, 133–34, 145, 148, 152–53, 156,
  168; in *Romantic Encounters at the
  Dome*, 26, 133–34, 158, 162, 168
Custer's Revenge, 141–42, 146

dating, 140, 144; apps, 176; computer,
  61; computers as substitute for, 132;
  games, 29; simulations, 177
daycare crisis, 22, 190n69
desires, 167, 176; childcare and, 100;
  child-rearing and, 26; children's,
  117; conflicting, 45, 57; dolls and,
  111, 117, 124; girlhood and, 128;
  individual, 23, 31, 36, 45, 53, 58;
  men's, 58, 72–73, 144; men's sexual,
  40; sexual, 24, 36, 42, 57, 60–61, 173;
  women's, 46–47, 49–50, 58; women's
  sexual, 34, 40, 49
digital media, 2, 13, 27, 175–76,
  178–79
digital signal processing, 10, 105,
  117
digital technology, 23; masculinity
  and, 139; in talking dolls, 111–12,
  114, 124, 126
D'Ignazio, Fred, 85, 95, 107
divorce rates, 21, 33, 36
doll makers, 101, 123

doll play, 7, 110, 113–14, 119, 121–22; histories of, 174; traditions of, 26, 102, 130

dolls, 9, 26, 99–100, 111, 121–22, 201, 203–5; Black, 123; microprocessor enabled, 6; as proxies, 6, 26, 100–101, 103, 109, 111–12, 116, 121, 123–24, 128, 172; socialization and, 100–101, 110–11, 122, 174; sound activation and, 100, 112, 114–15, 121. *See also* talking dolls

domestic culture, 2–3, 6, 8, 65, 172, 174

domesticity, 3, 7, 92, 102, 128, 130, 172, 180; masculine, 64–65, 71–72, 77, 97; middle-class, 177

domestic life: computing and, 6, 17; digital media and, 27; men and, 25, 64, 71–72; robots and, 25, 72, 79, 81; women and, 66

domestic sphere, 23, 66

dominance, 11, 57, 132; masculine, 147; over women, 167

domination: masculine, 167; masculine sexuality and, 142; masculinity and, 132, 138–39; sex and, 151

Drucker, Donna J., 61–62

Ehrenreich, Barbara, 34, 43, 69, 137

Emerson, Lori, 7–8

Ensmenger, Nathan, 92, 174

everyday life, 10, 23, 92; computerization of, 64; computing and, 8–9, 13, 20, 73; robots and, 73, 82

family, 3, 11–13, 17, 19–21, 26–27, 37, 66–68, 92; bonds, 19; heteronormative, 6; labor and, 35, 66, 71; middle-class, 23, 86, 133; New Men and, 137; nuclear, 168; robots and, 63–65, 72–82, 84–86, 91, 95, 97, 176, 199n75; routines, 16, 25; software industry

and, 40; video games and, 129, 134, 146, 160; wage, 22. *See also* companionate family; family life; family relations; family relationships

Family Computing, 19–21, 32, 36, 70, 85, 105

family life, 13, 35, 69; companionate, 20–21, 65, 82, 92, 97; companionate computing and, 24; computers and, 2, 4, 6–8, 27, 68; computing and, 19–20; everyday practices of, 17; men and, 23, 70 (*see also* participant fatherhood); middle-class, 11, 64; robots and, 25, 64–65, 73–74, 77, 80, 97; work and, 66

family relations, 2, 15, 69; companionate, 3, 6, 14, 86, 172

family relationships, 2, 5 (fig.), 7, 11, 23, 65, 67, 78, 146, 175; companionate, 17

fatherhood, 22–23, 64, 68, 71, 97; classes, 87; companionate, 74; discourses about, 93; norms for, 70; robots and, 74. *See also* participant fatherhood

Fatherhood Project, 68, 70

femininity: girlhood, 120; hyperfemininity, 127; ideal, 110; obedient, 122; white girlhood, 112, 123

feminism, 40, 70; girl power, 122; new masculinity and, 137. *See also* antipornography feminists

feminist critique, 3, 26, 31, 49, 65, 133, 145, 151; of companionate relations, 175; of the couple, 32; of masculinity, 156

feminist culture, 3, 101, 180

feminist media histories, 3, 174

feminist men's movement, 68, 70

Frantz, Gene A., 105, 117, 200n1

Friends or Lovers, 31, 49, 59, 193n42

Frisina, Tom, 82, 90

Galoob, 110–16, 121, 123–26, 128, 175. *See also* Baby Talk

gender, 12, 109, 127, 129, 141, 177, 180; gap, 18, 102, 107; identity of robots, 88; as infrastructure, 174; *Interlude* and, 46; norms, 136, 168; participant fathers and, 70; relations, 7, 31, 137, 140, 148, 172; *Romantic Encounters at the Dome* and, 158

gender roles, 33, 41, 51, 56, 61, 70–71, 129, 156; changing expectations for, 37; shifting, 50; traditional, 21, 60, 86

General Electric, 13–14, 85

General Motors (GM), 68, 84, 86

Giddens, Anthony, 36, 60

girlhood, 100–102, 110–12, 114, 120–23, 128–30; neoliberal, 101, 122; white, 112, 122–23, 203n37

girls, 12, 18, 105, 107–8, 127–28, 212; computing and, 129–30; dolls and, 6, 26, 99–102, 109–17, 119–26, 130, 174, 204n66, 205n70; girls' game movement, 129, 206n88; Logo and, 107

Gore, Tipper, 147, 174

hackers, 3, 7, 136

hardware, 6–7, 16, 40, 92, 109, 132; configuration, 10; development of, 3; executives, 4; producers/makers, 2, 6, 11, 13, 19, 172; women and production of, 174

Havas, Julia, 133, 148

Hay, James, 9, 23

Heath Company, 25, 63–64, 73–82, 84, 88, 93; Heathkits, 75, 91

Hero Jr., 63–64, 72, 74, 79–81, 87–88, 90–91, 93, 176; programming manual, 199n74

Hero 1, 72–79, 87, 90, 93–96

Hess, Elizabeth, 34, 43

heterosexuality, 134, 136, 140, 155; games and, 167–68 (see also *Leisure Suit Larry in the Land of the Lounge Lizards*); masculine, 26; masculinity and, 133, 137, 156, 168; monogamous, 145

heterosexual relationships, 11, 37, 62, 134, 136–37, 167; fulfilling, 41; gender roles and, 136, 156; masculinity and, 132

Hicks, Mar, 61, 174

Hite Report, The, 43, 135

hobbyists, 2–4, 7–10, 15–16, 65, 76, 97; Altair and, 188n37; Hero 1 and, 74, 79; magazines and, 72; male, 38, 59, 64, 83, 87, 89; participant fatherhood and, 96; *Softporn* and, 142

Hochschild, Arlie, 22, 35, 69, 97

Homebrew Computer Club, 15, 188n39

home computers, 3–4, 12, 16–20, 171; as domestic appliances, 8; market for, 38, 52, 79, 175; personal robots and, 63–64; as product category, 186n5; romance software and, 29, 31, 41, 47. *See also* Apple II; Commodore: PET; Commodore: TRS-80; Texas Instruments (TI)

home computing, 2–4, 8, 12, 16–18, 64, 135

hothousing, 104–7, 118, 125–26. *See also* superbabies

How to Make Love to a Computer, 135–37

Hubot, 72–74, 83–85, 88, 93, 176

Hubotics, 25, 63–64, 73, 83, 88; Hubot, 72

IBM, 14, 16; PC, 3, 131

identity, 51, 138, 154, 165, 168; gender, 88; masculine, 11; sexual, 42

inCider, 40, 48 (fig.)

individualism, 167; masculine, 11, 32; rational, 165; rugged, 108

information economy, 22, 68–69, 74, 122
information technology, 60–61
InfoWorld, 47, 71, 90 (fig.), 102–3
insecurity, 11, 69, 164–65
interaction, 10, 12, 125–26; computer, 9, 46, 65, 97, 109, 115–17, 128–29 (*see also* Baby Talk; Julie; talking dolls: computer); in Coral, 178; cybernetic, 205n70; rational style of, 5; robot, 87, 93, 97; sexual, 6; social, 124; between users, 31; with video game characters, 162–63
Interlude, 24, 31, 37–49, 52–53, 57–58, 175, 178–79, 193n29
Interlude II, 47, 48 (fig.), 60, 193n36
interpersonal relationships, 132, 136, 161
intimacy, 6, 23–24, 34–35, 41–42, 139; companionate, 25; computing, 95; computing and, 32; coupled, 7, 31–32, 47; discourses, 31, 34, 41, 52; *IntraCourse* and, 55–57; *Lovers or Strangers* and, 52; marital, 37; normative conceptions of, 61; nurturing, 50; physical, 45, 51; romance software and, 58; *Romantic Encounters at the Dome* and, 161, 163; sexual, 36, 132
Intracorp, 53–56, 59, 61
IntraCourse, 24, 31, 37–38, 40–41, 52–61, 173, 175
iRobot Roomba, 13, 176

Jacobs, Gloria, 34, 43
Jibo, 86, 89
Jill, 100, 108, 110, 112, 119–25, 127, 129, 177; cybernetic interaction and, 205n70; sales of, 175
Julie, 99–100, 106, 108–12, 116–27, 129, 177; cybernetic interaction and, 205n70; sales of, 175; speech

recognition and, 99–100, 109, 111–12, 116–17, 119–20, 125, 200n1
Joy of Sex, The (Comfort), 34, 53, 135

Kaplan, Danny, 167, 177
Kinsey, Alfred, 55; reports, 62, 194n47
Kirkland, Ewan, 167, 207n6
kit computers, 4, 188n37

labor, 35, 52, 66, 70; automated, 88; clerical, 174; computer, 92–93; computers and, 67, 139; couples and, 33; domestic, 71–73, 82, 95–97; men's, 68, 73; programming, 65–66, 74, 92, 96–98; robots and, 75–77, 86, 94, 98; women and, 22, 69, 82, 84, 103–4. *See also* childcare
La Ruina, Richard, 177–78
Leather Goddesses of Phobos, 142, 147
Leisure Suit Larry in the Land of the Lounge Lizards, 26, 133–34, 139, 141–58, 168, 174–75, 177–78. *See also* cringe; cringe aesthetics; masculinity
Leppert, Alice, 71–72, 104, 190n69
Levine, James A., 67–68
Light, Jen, 92, 174
Logo, 67, 103, 105, 107–8
Lovers or Strangers, 24, 29–31, 37–38, 40–41, 47, 49–53, 57–60, 173, 175, 179. See also *Friends or Lovers*
Lowe, Al, 144, 146–47, 150–54, 157. See also *Leisure Suit Larry in the Land of the Lounge Lizards*

machine cuteness, 88–89. *See also* robots: cuteness and
MacPlaymate, 143–44
Magnavox Odyssey, 128, 206n82
Maher, Jimmy, 154–55
mainframes, 13–14, 16
marital relationships, 33–34

marketing, 59, 142; of computers, 79; of digital devices, 13; of dolls, 118, 121, 123; to girls, 101; for *IntraCourse*, 55; for *Leisure Suit Larry in the Land of the Lounge Lizards*, 144–45; for *Lovers or Strangers*, 47, 50, 173; niche, 127; of robots, 25, 65, 69, 76, 84, 86, 97; romance software and, 31

marriage, 23, 31–37, 61; bonds, 21; counseling, 59–60; heterosexuality authorized by, 156; in *Leisure Suit Larry in the Land of the Lounge Lizards*, 149, 151; *Lovers or Strangers* and, 59; masculinity and, 137; in *Romantic Encounters at the Dome*, 166

masculinity, 86, 131–34, 136–40, 169; computing and, 92; critiques of, 26; domestic, 71, 77; failed, 178; heteromasculinity, 6, 133–34, 139, 155; heterosexual, 139, 154, 158, 168; juvenile, 88; *Leisure Suit Larry in the Land of the Lounge Lizards* and, 144–45, 147–51, 153–58; *MacPlaymate* and, 143–44; new, 137–38, 158, 162, 166; norms of, 23, 68, 70, 140; *Romantic Encounters at the Dome* and, 158–59, 162–66; video games and, 102, 134–35, 167–68

Masters and Johnson, 55–56

mastery, 10–12, 77, 132, 134, 137, 167; technological, 139

masturbation, 53; women's, 33, 56

maternal care, 26, 103, 111, 172

Mattel, 110, 123; Teach and Learn Computer, 106, 109

media, 174–75, 178; adult, 133, 140–41, 147, 150, 174; childhood and, 127; computers as, 2, 6, 11, 24, 172; culture, 132; digital, 2, 13, 27, 175–76, 178–79; feminist, 3; historians, 66, 85, 126; interactive, 143; popular, 71;

pornographic, 138, 140, 144–45, 150–51; producers, 126; proxies, 101; relations, 106; social, 212n18

men, 2–3, 5, 12, 18–19, 23–26, 31, 40, 68–72, 135–38, 190, 196, 208, 210, 212; adult games and, 134, 139–40, 142–44, 146, 150–51, 166; heterosexual, 38, 42; New Men, 132–33, 137, 156, 158, 162, 166, 208n17; programming and, 92; robots and, 64–66, 72–77, 79–82, 84–87, 90, 94–95, 97, 173, 176–77; romance software and, 35, 42, 44, 46–47, 49, 52, 59, 61; in the seduction community, 167. *See also* fatherhood; masculinity; participant fatherhood; wormboys

microcomputers, 8–11, 15, 38, 63–64, 102, 105; complaints about, 47; market for, 25

Microillusions, 131–32, 158–60. See also *Romantic Encounters at the Dome*

microprocessors, 9, 15, 25, 93, 100, 109–11, 115–16; history of, 188n36. *See also* talking dolls

microprocessor technology, 9, 14, 115; dolls and, 111–12, 120; personal robots and, 25, 63, 73

middle class, 21, 69, 104, 137, 156

middle-class lifestyle, 22, 69

monogamy, 58–59

motherhood, 22, 101–2, 128

mothers, 22, 26, 104–5, 127; computer talking dolls and, 100–101, 111, 116, 122, 124; Mr. Mom characters and, 72. *See also* supermoms

Newlywed Game, The, 50–52

Newman, Michael Z., 110, 129

Nooney, Laine, 8, 141, 174

Olsen, Brigitta, 49–50, 52. See also *Lovers or Strangers*

Papert, Seymour, 108

parenting, 68, 70, 86, 96, 103–4, 123–24; dolls and, 100

parents, 21, 103–6, 108–9, 127; anxiety and, 104; computers and, 5, 20; dolls and, 26, 100–101, 111, 114, 116, 118–26; robots and, 79, 82, 177. *See also* boys; children; girls

Parents, 66, 68

participant fatherhood, 22, 25, 64–65, 70–71, 81, 85–87; computers and, 24; expectations of, 76, 80, 84, 97; performances of, 6; robots and, 73–74, 94–96, 172

passion, 23, 31, 34–35, 42, 92

Patton, Elizabeth, 25, 66

Penley, Constance, 151

personal computers, 8, 15–16, 23, 37, 92, 136; as commodities, 10, 12; cultures, 139; Heathkits and, 75; microprocessor technology and, 14; personal robots and, 25, 63; romance software and, 31–32, 58; at school, 20; strip games and, 143; Topo and, 81; women and, 66

personal computing, 2, 8–11, 14–16, 32, 38, 92, 188n32, 188n39

personal robots, 25, 63–65, 69, 72–74, 84, 92–93, 95, 173; cuteness and, 89, 91, 93; masculinity and, 88; men and, 97; parenting and, 86; women and, 82. *See also* Androbot; Heath Company; Hubotics

phonography, 3, 127, 174

play, 68, 96, 106–7, 109, 111, 124, 126–27, 177; adult games and, 60, 143, 148–49, 162, 165; boys' cultures of, 102; computer, 26, 128, 130; doll, 7, 26, 102, 110, 112–16, 119–22, 124–26, 130, 174; feminine forms of, 101; girlhood, 100, 111, 114, 130; girls', 110–12, 117, 124–25,

128; sexual, 41–44, 46, 53; women and, 84

Playmates, 110, 119–21, 123, 175. *See also* Jill

Popular Computing, 29, 78

Popular Mechanics, 78, 94

pornography/porn, 43, 138, 140–41, 143, 174; adult games and, 133, 140, 142, 143–45, 147–48, 150–51; couples, 23; domesticated, 34; masculinity in, 151. See also *Softporn Adventure*

professional development, 25, 81

programming, 9, 63–64, 67, 69, 79, 83, 85, 87, 93–94; children and, 105, 107; children's television, 127–28; dolls and, 109, 125 (*see also* Julie); gendering of, 93–92; heterosexual male culture of, 46; labor, 65–66, 96–97; languages, 67, 103; manual for Hero Jr., 199n74; men and, 25, 64–66, 72, 74–77, 86 (*see also* Hero 1); profession, 174. *See also* Hubot; robots; Topo

psychologists, 4, 19, 55, 59, 106, 173; child, 113

public life, 21, 173

pure relationships, 36, 60

radio, 3, 55, 83, 93, 174; ham, 75; signals, 200n89

RadioShack, 15, 47

Rankin, Joy Lisi, 12, 14

rape, 138, 142

Raskin, Robin, 36–37

rationality, 158, 164; avatar, 134; masculine, 157, 164–65

RB5X, 73, 85

Reed, Lori, 6, 18, 186n9, 191n2

relationality, 169; caring, 89; children and, 65; companionate, 9, 19, 24, 27, 39, 134, 175; coupled, 41; domestic,

3, 6, 12, 24–26, 135, 172, 180; heterosexual, 133–35; romantic, 62

relations, 21; companionate, 11–12, 41, 55, 85, 174–75; to computers, 136–37; domestic, 6; gender, 7, 31, 137, 140, 148, 172; sexual, 33–34, 42, 55, 60, 158; social, 11; with women, 131, 138, 151, 160, 172. *See also* family relations

relationships. *See* companionate relationships; coupled relationships; domestic life; family relationships; heterosexual relationships; interpersonal relationships; marital relationships; pure relationships; romantic relationships; sexual relationships; social relationships

roboticists, 13; amateur, 72–73

robotics, 13, 63, 73, 75, 81, 97; consumer, 72; fathers and, 65; home, 73–74, 76–77; industrial, 68, 75; labor, 94, 98; men and, 25, 72, 74, 77, 79–80, 84, 86; training, 69, 77. *See also* Androbot; Heath Company; iRobot Roomba

Robotics Age, 72, 95

robot manufacturers, 64, 72, 86–87, 176. *See also* Androbot; Heath Company; Hubotics

robots, 6, 9, 13, 25–26, 63, 69, 81–85, 95, 134, 175–77; as appliances, 81, 83, 173; childcare and, 64–67, 72–75, 78, 80–82, 84–88, 96–97, 176; as children, 94, 96, 176; cuteness and, 87–93, 97, 173, 199n70; family and, 79–80, 86; family life and, 25, 64–65, 73–74, 77, 80, 97; industrial, 68, 72, 75–77; men and, 64–66, 72–77, 79–82, 84–87, 90, 94–95, 97, 173, 176–77; participant fatherhood and, 73–74, 94–96, 172; women and, 25, 64–65, 82–86. *See also* Androbot;

BoB; Hero 1; Hero Jr.; Hubot; Jibo; personal robots; RB5X; Topo

romance, 24, 31–35, 37, 39, 41, 44, 50, 58, 61, 132, 135–36, 155, 176; game, 168

romance novels, 50; aesthetics of, 50, 173; covers of, 29, 30 (fig.), 40, 47

romance software, 24–25, 31–33, 35, 37–41, 47, 58–62, 134, 172–73, 176, 178–79; women's desires and, 49. See also *IntraCourse*; *Lovers or Strangers*

Romantic Encounters at the Dome, 26, 131–34, 143, 158–66, 168, 177. *See also* masculinity

romantic relationships, 23–24, 35–36, 41, 132–33, 176

Rotundo, E. Anthony, 22, 70

Ruberg, Bo, 149, 168

Salter, Anastasia, 155, 167, 207n6, 210n69

Sears, 17; Home Control System, 19; toy computers, 106, 109

seduction, 26, 32, 132, 140, 160, 163, 179; community, 167, 177–78; heteromasculine, 24; heterosexual, 6, 26, 166, 169; of women, 148

Seiter, Ellen, 127–28

self-help, 31; books, 34, 44, 135; culture, 7, 25; literature, 35, 58; sexual, 53; software, 19

sensitivity, 114, 132, 161, 163–64; emotional, 44; of Julie, 117; masculine, 132; of the New Man, 137; to women, 163

Sensuous Woman, The (Garrity), 34, 135

sex, 29, 32–34, 39–41, 49, 141, 144–47, 149, 151, 156–57, 161, 163–64, 166; advice, 40, 44; anonymous, 131; drive, 57; high tech, 59; lives, 31, 33, 36, 41–42, 44, 53; questionnaires, 176; software for, 38, 43, 45–46 (*see also* romance software); toys, 34, 143. See also *Leisure Suit Larry in the Land of the Lounge Lizards*; *Romantic Encounters at the Dome*

sex manuals, 23, 33–34, 43, 53, 54 (fig.), 135, 137

sex research, 53, 55–56, 58, 173

sexual competence, 145; masculine, 151

sexual culture, 33–34, 173

sexual fulfillment, 21, 24, 33–36, 136

sexuality, 33, 35, 41, 43, 49, 61–62, 168, 174, 179; adult games and, 133; companionate, 25; computing and, 32, 132, 141; domestic, 38; *Interlude* and, 43–44; *IntraCourse* and, 53, 55–57, 173; *Lovers or Strangers* and, 50–51; masculine, 142; misogynist, 146; romance software and, 40, 58, 60; swinging, 155; TV and, 140; violent male, 138; women's, 34; young teens and, 106. See also heterosexuality

sexual relating, 31, 34, 55

sexual relationships, 33, 35–36, 38, 58, 135, 165, 178

sexual revolution, 23, 34, 38, 135–36, 138; cultures, 140, 173

Shaw, Adrienne, 151, 154–55

Sidley, J. D., 32–33, 37

Sierra, 141, 144–48, 150–53, 158. See also *Leisure Suit Larry in the Land of the Lounge Lizards*; *Softporn Adventure*; Williams, John

smart technologies, 2, 13, 23, 85–86, 115–16, 176. *See also* Amazon Echo; iRobot Roomba

socialization, 110–11, 122; girls', 174; feminine, 100, 111; maternal, 101

social relationships, 11, 26, 134

Softalk, 29, 141

Softporn Adventure, 141–42, 145, 148, 150, 208n30

software, 3, 8–9, 11, 18–20, 24, 85, 109, 132, 142, 158–59, 174, 182, 193n42; adult, 145, 147; advertising, 50; children's, 112; companies, 19, 24, 41, 61, 133, 135, 139–40; compatibility, 57; couples, 44, 47; development professionals, 97; educational, 18, 20, 102, 105, 107, 128; firms, 18, 49; for girls, 108; home office, 17, 29; industry, 18, 40, 49–50, 141–42, 145; masculine culture of, 42; packaging, 55; producers/makers, 2, 6–7, 13, 53, 135, 172; publishers, 141, 144–46, 160; recipe, 96, 172; for sex and romance, 38; spreadsheet, 10, 16, 70; word processing, 16, 189n46. *See also* romance software

Spigel, Lynn, 84, 185n4

spokespeople, 147, 171, 208n17; Androbot, 82; celebrity, 17. *See also* Alda, Alan; Cavett, Dick

spouses, 18, 21, 23, 33, 37

spreadsheets, 7, 16

Streeter, Thomas, 10–11

Sulimma, Maria, 133, 148

superbabies, 22, 125–26; burnout and, 107

supermoms, 22, 26, 104, 202n14

Syntonic, 38, 40, 42, 46–47. See also *Interlude*

talking dolls, 9, 99, 108, 114, 116, 123, 130, 177; anxieties about, 205n73; computer, 26, 100–103, 108–12, 122, 124–30, 173; mechanisms, 204n50; neoliberal girlhood and, 122; play, 121. *See also* Baby Talk; Jill; Julie

telecommuting, 66, 70

television (TV), 2–3, 42, 71, 83, 129, 140, 146, 174; audiences, 51; Brothers and, 55; children's programming for, 127; commercials, 83, 120, 171; consumption, 125; dramedy, 148; as family medium, 17, 33; game shows, 29, 50–51; network, 140; representations of men on, 71–72; smart TVs, 13, 23; toy–TV crossovers, 127; women's genres of, 128

Texas Instruments (TI), 16–17, 105–6, 117, 189n50

Thomas, Lee, 158, 160

Thomas, Veronica, 159–60

Topo, 72–74, 81–85, 87–90, 93–94, 176, 200n89

toy companies, 73, 99, 108–9, 111–12, 119, 123, 173. *See also* Coleco; Galoob; Mattel; Playmates; Worlds of Wonder

toy makers, 110–11, 129; gender and, 127

toys, 4, 113, 123, 125–27, 129–30, 175; anthropomorphic, 100, 110; computer, 106, 109; electronic, 105, 109–10; for girls, 108, 122; microprocessors and, 9, 111; robots as, 73; sex, 33–34, 143 (*see also* vibrators); talking, 124, 126. *See also* Baby Talk; dolls; Jill; Julie; Worlds of Wonder: Teddy Ruxpin

Turkle, Sherry, 3, 136, 203n37

Turner, Fred, 10, 14

unemployment, 22, 69, 76; technological, 75, 88

VCRs, 23, 115, 129

vibrators, 23, 33, 43–45

video game consoles, 9, 16, 83, 128, 141, 206n82

video games, 4, 6–7, 9, 18, 20, 37, 67, 102, 109–10, 127–30, 206n82, 206n88; action-adventure, 29; adult, 26, 39, 41, 132–35, 139–46, 149, 154, 158, 167–69, 172; dating and, 177; flight simulator, 10, 18; text-based adventure, 39, 141 (see also *Softporn Adventure*). See also *Custer's Revenge*; *Interlude*; *IntraCourse*; *Leisure Suit Larry in the Land of the Lounge Lizards*; romance software

violence, 146; game, 67; gendered, 167; sexual, 138, 144

VisiCalc, 10, 16–17, 19, 32

Vora, Kalindi, 76, 86, 177

White, Mimi, 51–52

Williams, John, 144–47. See also *Leisure Suit Larry in the Land of the Lounge Lizards*; *Softporn Adventure*

women, 3, 12, 17–18, 21, 32–37, 46, 50, 52, 66, 129, 172–75; adult games and, 26, 131–39, 141–52, 156, 158, 160, 162–69, 177–78; Brothers and, 55; with children, 20, 22, 103; computer science and, 12, 92; homework and, 22, 35; robot, 199n75; robots and, 25, 64–65, 82–86; romance software and, 59; sexual desires of, 34, 40, 49, 58; working women, 22, 72, 100; work outside the home and, 21, 69, 71–72, 76, 84–85, 101, 103–4

women's culture, 31–32, 40, 49–50, 55, 61, 173

women's liberation, 21, 68, 136, 151

women's magazines, 18, 35, 40, 43–44, 105

word processing, 16, 67, 105, 189n46

Worlds of Wonder, 99–100, 106, 110, 112, 117–19, 121–22; Teddy Ruxpin, 99, 110, 121, 179, 203n43. *See also* Julie

wormboys, 132, 137–38

REEM HILU is assistant professor of film and media studies at Washington University in St. Louis.

Printed in the USA
CPSIA information can be obtained
at www.ICGtesting.com
CBHW051952061124
17024CB00002B/3